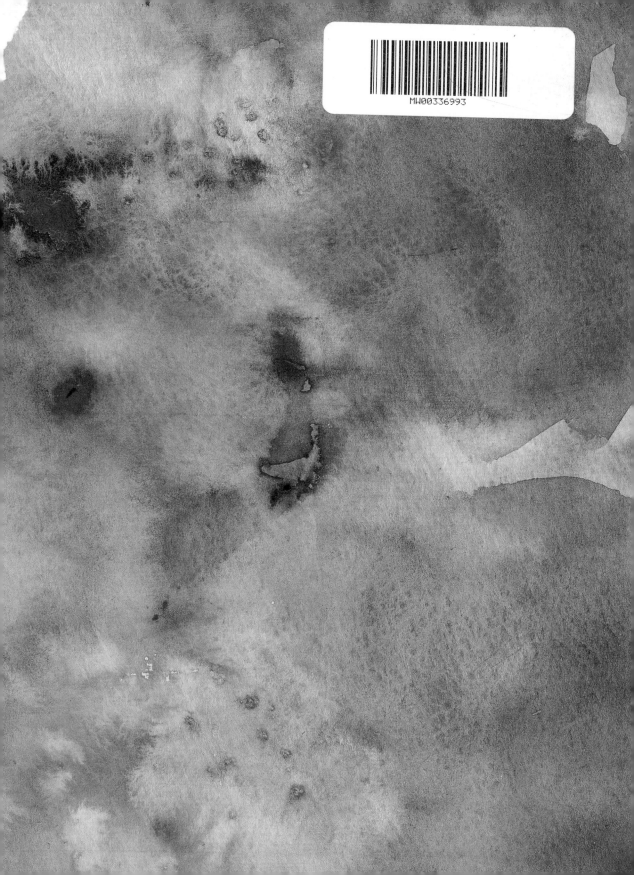

The Scandinavian Skincare Bible

Johanna Gillbro, PhD, is an award-winning skin scientist with more than fifteen years of experience in experimental dermatology, clinical research, and skincare product development, as well as substantial experience within the pharmaceutical industry. Gillbro is frequently engaged as a speaker at international dermatological and cosmetic science conferences to present her cutting-edge research, and for the past decade has been the most cited author in the *International Journal of Cosmetic Science*.

Fiona Graham has a degree in Modern Languages from Oxford University, and has lived in Kenya, Germany, the Netherlands, Luxembourg, Nicaragua, and Belgium. She translates from Spanish, French, Dutch, Swedish, and German, and is currently the reviews editor at the *Swedish Book Review*.

The Scandinavian Skincare Bible

the definitive guide to understanding your skin

Johanna Gillbro

SCRIBE

Melbourne • London

Scribe Publications

18–20 Edward St, Brunswick, Victoria 3056, Australia
2 John St, Clerkenwell, London, WC1N 2ES, United Kingdom
3754 Pleasant Ave, Suite 100, Minneapolis, Minnesota 55409, USA

Published by Scribe in 2020
Published by agreement with Ahlander Agency

First published in Swedish by Bookmark Förlag as *Hudbibeln* in 2019

The advice provided in this book has been carefully considered and checked by the author. It should not, however, be regarded as a substitute for individual medical advice. The author, their representatives, and the publisher shall not bear any liability whatsoever for personal injury, property damage, and financial losses.

Text design: Johan Barrett & Scribe
Photographs on pages ii, 8–9, 46–47, 50, 108–109, 138–139, 184: Anders Kylberg/Acne, Stylist: Emilie Florin/Acne
Photograph of the author, page 5: Annika Berglund
Watercolours on pages 14, 16–17, 19, 20, 81, 112, 130, 163: Ina Schuppe Koistinen
Illustrations: Olof Bandh
Images on pages 33, 78, 84, 107, 172, 189: Shutterstock
Photographs on pages 23, 25, 27, 34, 38, 43, 49, 65, 74, 82, 87, 99, 111, 116, 119, 122, 133, 141, 142, 153, 155, 166, 171, 179, 183, 186, 190, 194: Pexels
Photographs on pages 137, 165, 168, 180, 182: Unsplash
Photograph on page 69: Wikipedia
Photographs on pages 35 & 60: Author's own
Photographs on pages 36, 88, 101: Stocksy
Images on cover: J.D.S./Shutterstock

Printed and bound in China by 1010 Printing Co Ltd

Scribe Publications is committed to the sustainable use of natural resources and the use of paper products made responsibly from those resources.

9781925849851 (Australian edition)
9781912854943 (UK edition)
9781950354351 (US edition)

Catalogue records for this book are available from the National Library of Australia and the British Library.

scribepublications.com.au
scribepublications.co.uk
scribepublications.com

Contents

Foreword

We get goosebumps if someone strokes our skin gently. The reason for this is that a tiny hair-erector muscle — the *arrector pili*, located deep in each hair follicle — responds to nerve signals from the brain. Our hairs stand on end, and the layer of air they trap protects us against the cold.

The skin, our body's outer shell, is a fascinating organ that embodies memories from its evolutionary development. Yet given that it's our largest organ, with a surface area of two square metres, we know surprisingly little about it — either about its structure, or about how best to take care of it. We know that oily fish are good for the brain, that olive oil is good for the heart, and that the calcium in yoghurt strengthens the skeleton.

But what about our skin?

This book takes a fresh look at the skin from a holistic perspective.

What can the latest research tell us about our skin? What's your own personal skin type, and what is its current status? What can you do to make your skin healthier? And what role do the skin's micro-organisms play? How can nutrition and lifestyle affect skin health?

Science is advancing all the time, and right now major discoveries are being made about the skin's own microbiome: its bacterial flora, comprising hundreds of millions of micro-organisms that work together with skin cells and which, we now know, play a decisive role in our health.

At the same time, we expose our skin to too many antibacterial substances in our daily skincare products. Currently, scientific congresses worldwide are discussing how cosmetics and skincare affect our skin's microbiome, which is comparable to the gut microbiome, long known to be vulnerable to the effects of preservatives and processed food.

There's a plethora
of products — but often
no research to show
they do what it says on
the tin. Shouldn't we hit
the reset button?

To say that I find skin fascinating would be an understatement. My personal interest began in my childhood, when I had vitiligo, a disorder affecting the skin's pigmentation, resulting in patches or spots of lighter skin. I wasn't at all bothered by this as a child. I felt I was unique, and I was happy to identify with my favourite breed of dogs, dalmatians. I was proud of my patchy skin. But things got worse in my teenage years. The desire to be the same as everyone else replaced the desire to be unique.

Our skin affects our self-image. Some variations in our skin's appearance can lead to social stigmatisation. Apart from vitiligo, these variations include acne, psoriasis, rosacea, eczema, and melasma. These conditions can also significantly affect our mental health.

I consulted many dermatologists, and tried everything from cortico-steroids and UVB therapy to transplanting healthy skin cells with normal pigmentation to the affected areas in an effort to cure my vitiligo. Nothing worked. If anything, these treatments only made matters worse.

I began studying pharmacy at the University of Uppsala, where I graduated in 2002. My studies brought me into contact with Karin Schallreuter, the German dermatologist and skin scientist, who was a professor at the University of Bradford and an expert in vitiligo. Thanks to this contact, I had the opportunity to pursue doctoral studies in experimental and clinical dermatology as part of Dr Schallreuter's team. In 2006 I defended my thesis on vitiligo, graduating with a doctorate in experimental dermatology.

These days I'm at peace with my vitiligo, and I'm once again fond of my white patches, although many have now disappeared completely. If it hadn't been for them, I might not have had the same drive, experienced all that I have, or met all the remarkable scientists and other people I've encountered over the years.

My expertise in skin-related matters led in due course to the skincare industry, and today I have over fifteen years' experience of skincare in a variety of contexts. Within the industry, I've worked as a research team leader and as a head of innovation, responsible for developing new active ingredients in skincare products that really work. During my years in this sector, I've realised that the general public know relatively little about it. Do you understand the structure of a skin cream or what ingredients it contains?

There's a plethora of products, but often there's no research to show whether they have the desired effect. Conversely, using too many products can actually worsen the skin's condition.

So shouldn't we hit the reset button?

And if we're going to start over, which products work, and which ones are superfluous?

I knew writing this book would create certain expectations, and I hesitated for a while before taking the plunge. One reason was that as a scientist, I'm not often in a position to give people what many of them want — simple, straightforward answers. Which skin cream is best for dry skin? Does collagen really work? What do you think about serum?

In most cases it's hard to give simple answers. In fact, there may be only three things that really work: commitment, knowledge, and a holistic perspective. I have taken these as my guiding principles in writing *The Scandinavian Skincare Bible.*

Johanna Gillbro

Dermatology — A Crash Course

You'll be meeting many new words and concepts in this book. We'll start with a list of key terms that it will be handy to familiarise yourself with from the outset.

THE BARRIER FUNCTION

The outermost layer of the skin, called the *stratum corneum* (Latin for 'horny layer'), is made up of flat cells covered by a film that retains moisture, known as the *hydrolipidic film*. The hydrolipidic film consists of *sebum*, water, and *humectants* (substances that retain moisture). Sebum is a blend of oils secreted by the sebaceous glands. The humectants are made up of salt and acids. The oils, salts, and acids in the hydrolipidic film occur naturally in the skin. This combination of substances, together with the stratum corneum, functions as a barrier. It's a formula that has yet to be matched by any skin cream.

WATER LOSS — TEWL

The evaporation of moisture from the skin is known as *transepidermal water loss* (*TEWL*). It involves the transfer of water through the outer layer of skin, the *epidermis*, into the surrounding atmosphere. This water evaporates more slowly from a young, healthy skin and faster from an ageing or dry skin. Applying skin cream prevents evaporation in two ways: the humectants retain moisture, and the oil in the cream holds in water by means of *occlusion*.

MICRO-ORGANISMS

Like the gut, the skin has a *microbiome*, comprising thousands of species of bacteria, fungi, and viruses. Most of these organisms are good for us: they protect us against *pathogenic* bacteria, help hydrate the skin, and have an anti-inflammatory function. The communication that takes place between the gut and skin microbiomes is called the *gut–skin axis*. Imbalances between gut and skin have been linked to complaints such as acne, psoriasis, eczema, and rosacea.

OXIDATIVE STRESS

Oxidative stress occurs naturally in all biological organisms that use oxygen, being an unwanted result of cellular respiration. *Free radicals*, which damage cell membranes if they're not broken down in good time, are a by-product of respiration. Our *antioxidant enzymes*, together with the *antioxidants* in our food, are the substances responsible for breaking down these free radicals.

AGEING

Both natural and external factors play a role in skin ageing. Ageing in general is known as *senescence*, a condition in which skin cells become more passive. As we get older, the body's ability to withstand oxidative stress lessens. Other factors, such as sunlight, cigarette smoke, and pollution, also contribute to oxidative stress in the skin, hastening the onset of senescence. Unlike the antioxidant enzymes that occur naturally in the skin, the levels of a group of enzymes known as *MMPs* rise with increasing age or when we are exposed to sunshine, pollution, or cigarette smoke. MMPs destroy the strings of *collagen* and *elastin* in the skin's other smart layer, the *dermis*, resulting in looser skin with more wrinkles.

SKINCARE AND DERMATOLOGY

Skincare products are a subsection of cosmetics, their main purpose being to cleanse or perfume the skin, alter people's appearance, provide protection, maintain the skin in good condition, or neutralise body odour. Cosmetic skincare products, however, are not suitable for treating skin complaints such as acne, eczema, rosacea, psoriasis, skin infections, or vitiligo. *Dermatology* is the science of the skin and the diseases that can affect it. Dermatological treatments are medical, designed to cure or relieve skin disorders.

DERMATOLOGY – A CRASH COURSE

About the Skin

The skin is our largest organ. With its surface area of two square metres, it covers us, providing protection against the sun, bacteria, and dangerous substances. It passes on and receives information through the scars and marks it bears, its colours, sensory cells, hairs, and receptors. It's time to rediscover our skin.

Three Smart Layers

Human skin is divided into three main layers: the epidermis (the outermost layer), the dermis (the layer directly beneath the epidermis), and the hypodermis (the innermost layer). Each layer in its turn is made up of further functional layers that together form an elastic mass of epidermal, dermal, and hypodermal cells, immune cells, blood vessels, sebaceous glands, and sweat glands.

These cells work together with the countless trillions of good bacteria and fungi that inhabit the skin. Together they determine the skin's health status. Some people have oily skin, others dry, while a smaller number have medical conditions. And all these aspects are subject to change, both with the seasons and with the passage of the years.

This chapter will examine the skin's structure in detail, help you get to know your skin status, and tell you about some of the most common skin complaints. We'll also take a closer look at the intrinsic and extrinsic reasons for skin ageing, and at how the sun affects the skin — both for bad and for good.

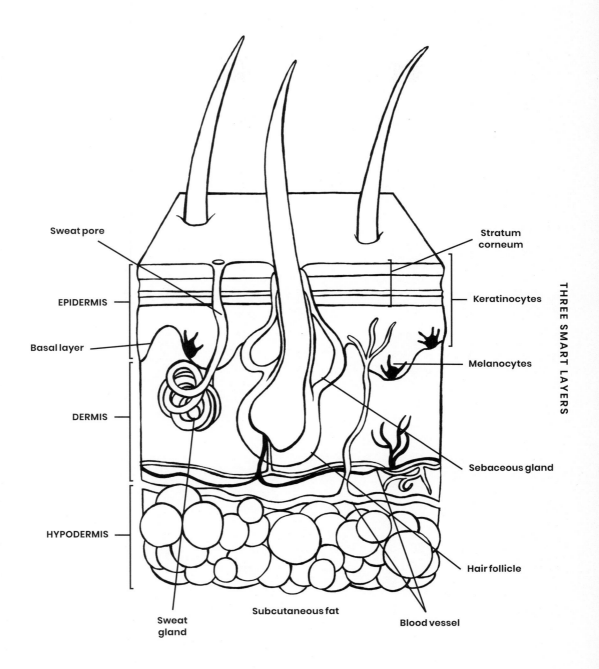

Sweat pore

Stratum
corneum

EPIDERMIS

Keratinocytes

THREE SMART LAYERS

Basal layer

Melanocytes

DERMIS

Sebaceous gland

HYPODERMIS

Hair follicle

Sweat
gland

Subcutaneous fat

Blood vessel

CROSS-SECTION OF THE SKIN

The skin's three main layers — epidermis, dermis, and hypodermis

The Epidermis — The Skin We Can See

The epidermis is the part of the skin that we can see and touch. It's only 0.05–0.1 millimetres thick in general, though slightly thicker on our soles and palms. The epidermis is covered by an emulsion of water and fats known as the hydrolipidic film. This comprises natural humectants such as salts and glycerine, together with a unique mixture of cholesterol, ceramides, free fatty acids from sebum, triglycerides, and squalene. The hydrolipidic film keeps the skin pliant and acts as a protective barrier against harmful bacteria and fungi. It is acidic, with a pH value between 4.7 and 5.5. The reason for its low pH value is the large numbers of good bacteria that inhabit the skin, which spend their time secreting acids (such as lactic acid) to fight off the bacteria that have no place there.

THE SKIN'S FIRST LINE OF DEFENCE

If a substance nonetheless manages to penetrate the skin — perhaps because the hydrolipidic film is depleted and its barrier function weakened — the epidermis has an army of vigilant Langerhans cells which are constantly on the lookout. Langerhans cells are not unlike an octopus in shape. Their outstretched tentacles capture foreign substances, such as allergens from cosmetics or perfume. The allergens are broken down in the

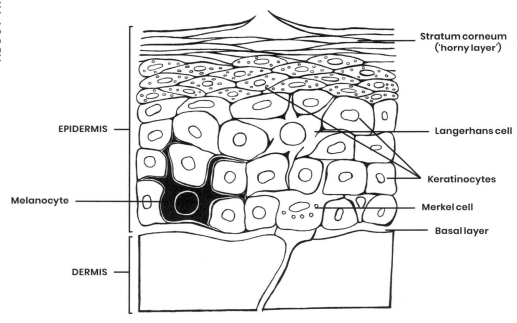

CROSS-SECTION OF THE EPIDERMIS

The epidermis is the uppermost layer of the skin, the layer that can be seen and touched. It is the skin's first outer line of defence.

Langerhans cells, and the resultant material is transported to the lymph nodes that form part of the immune defence system. These may react by sending more immune cells to the skin via the bloodstream.

RENEWAL AND PIGMENTATION

Have you heard it said that our skin renews itself every twenty-eight days? There's a lot of truth in that statement, but the change is gradual: the skin forms new cells and eliminates those that have done their job. The two main types of cell in the epidermis are *keratinocytes* and *melanocytes*. Keratinocytes, which are in the majority, are responsible for regenerating the epidermis. They are embedded in the unique hydrolipidic film that holds the cells of the epidermis together and protects them from dehydration.

The skin's regeneration results from the keratinocytes moving to the outer surface of the skin, the *stratum corneum*, where they lose their cell nuclei and die. The skin gradually sloughs off the dead cells; we replace several kilos of dead skin every year.

If this renewal process goes awry — if the skin's keratinocytes are attacked by its own immune system, for example — the result is the disorder known as psoriasis. What happens then is that the keratinocytes don't die, but start to reproduce and multiply. That's why the skin of psoriasis sufferers often displays thick plaques.

Melanocytes are pigment cells that produce the colour in our skin. They are far less numerous than keratinocytes, with about one melanocyte to every forty keratinocytes. Melanocytes are located in the deepest layer of the epidermis (the basal layer) and have just one function — to secrete pigment. They're equipped with arms known as dendrites, which channel the pigment they produce into the surrounding keratinocytes so as to create an even skin tone.

There are two types of pigment: *eumelanin* (black or brown pigment) and *pheomelanin* (red pigment). Everyone has a combination of these two types. The less eumelanin we have, the more we need to protect our skin cells against the sun, as it's eumelanin that provides a natural sunscreen in darker skin types. We don't yet understand why we have pheomelanin and what function it fulfils. What we do know is that the more pheomelanin a person has, the ruddier their skin.

Eumelanin is the best defence against the sun and thus against damage to DNA in skin cells, whereas pheomelanin can cause what's known as oxidative stress through UV radiation, in addition to the damage caused by sunburn. This makes it particularly important for people with pale or pink skin to protect themselves against strong sunlight, both to avoid premature ageing and to reduce the risk of cancer. We'll come back to the impact of oxidative stress on the skin soon.

THREE SMART LAYERS

ABOUT PIGMENT

Did you know that pigment fulfils several functions in the body? We have pigment in our hair — but also in our ears (where it plays an important role in hearing and balance), in our eyes, and even in our hearts and brains. The main function of pigment in the brain is to bind heavy metals and to protect against toxic damage to the nerve cells. Interestingly, a link has been found between low melanin levels and the relative severity of Parkinson's disease. The less melanin a Parkinson's patient has in their nerve cells, the more serious the disease.

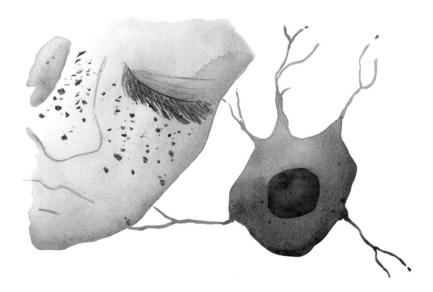

KERATINOCYTES (ABOVE)

Keratinocytes account for most of the cellular mass in the epidermis. They are tightly packed, forming a wall of tile-like cells. Keratinocytes are responsible for regenerating the epidermis.

MELANOCYTES (BELOW)

Melanocytes are pigment cells that give rise to the colour in our skin. They have arms known as dendrites, which channel the pigment into the surrounding keratinocytes, making for an even skin tone.

OUR SKIN PIGMENTATION

There's a reason for the pigment in our skin. Dark pigment shields the nucleus of a skin cell against potentially carcinogenic DNA damage, while also preserving the cell's ability to reproduce in the event of exposure to strong ultraviolet (UV) light. The colour of the skin is an evolutionary adaptation to the amount of solar radiation we're exposed to. But this takes time, a long time.

From about 1.2 million years ago to less than 100,000 years ago, we were all dark-skinned. It was only as humans moved northwards that lighter skin tones evolved. At some point, some northern populations experienced positive selection for lighter skin due to the fact that it is easier to produce sufficient amount of vitamin D with limited sunlight if you have lighter skin. The darker the skin, the more sun needed for vitamin D production, and since there are periods without sufficient sunlight in the northern hemisphere, the harder it is for darker skin-types to produce sufficient vitamin D levels.

To summarise: If people living at northerly latitudes had retained darker skin, they wouldn't have been able to produce enough vitamin D, which we need for a robust skeleton, our immune system, foetal development, and our mental health. Conversely, this means that people with darker skins living in northern countries may need to take vitamin D supplements during the coldest six months of the year.

But what about the opposite scenario: people migrating from the northern hemisphere to the southern? For example, Irish and British people moving to Australia in the 1800s, or the Dutch to South Africa and the Spanish to South America several centuries ago. Perhaps it sounds like a long time — a couple of centuries — but from an evolutionary perspective, it is like a drop in the ocean. There is too short time for skin colour to adapt to the higher exposure of UV light. Research by Nina Jablonski suggests that about 10,000 to 20,000 years is required for human populations to adapt their skin colour in a particular geographic area.

Unlike the darker-skinned people living in the northern hemisphere, individuals with fairer skin living in the southern hemisphere have no problem with the vitamin D levels, but instead the higher UV exposure means higher risk for UV damage, which leads to premature ageing of the skin, as well as a higher risk of developing skin cancer.

The most studied gene mutation, which is especially associated to fairer-skinned individuals, in particular, those with freckles and red hair, is the melanocortin receptor 1, which has a genetic predominance for skin cancer and especially melanoma. The consequence of this is that people with fairer skin living in the southern hemisphere need to have a sensible relationship to the sun.

Australia experiences some of the highest levels of UV radiation in the world because of the short distance to the equator. Approximately 60 per cent of the day's total carcinogenic radiation is received between 10am and 2pm — so stay out of the sun during this time, and protect yourself with clothing, hat, and sunscreen.

THREE SMART LAYERS

FRECKLES

Freckles are small brown or reddish-brown patches of melanin in the skin's topmost layer. In children, they're the result of a genetic predisposition. Though not present at birth, they emerge when the child is a few years old. Genetic predisposition to freckles is often associated with red or ginger hair.

Freckles can also be the result of exposure to strong sunlight through sunbathing or lying on a sunbed. When the sun's rays penetrate the skin, they activate the melanocytes, resulting in patches of pigment that darken and multiply over time.

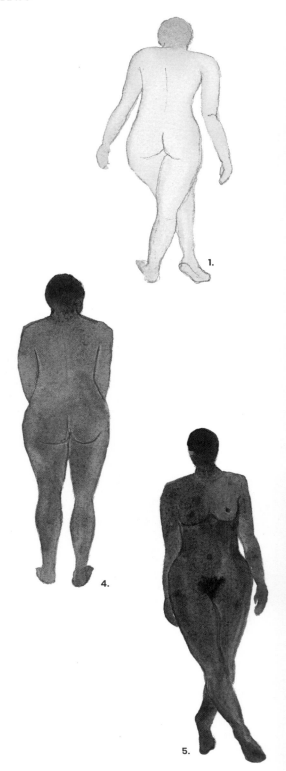

1.

4.

5.

WHAT HAPPENS WHEN WE HAVE LUNCH OUTSIDE? WHAT ABOUT THE DAY AFTER?

Melanocytes (pigment cells) react to sunlight. Exposure to UVA causes the pigment in the skin to darken straight away. If you have lunch out of doors, you'll see an almost immediate change in skin colour. This is because existing pigment is transferred from the melanocytes to the keratinocytes. UVB is responsible for the secretion of pigment, which usually takes twenty-four hours. This is the change in pigmentation you can see in the mirror the day after you've noticed you've caught the sun.

ABOUT THE SKIN

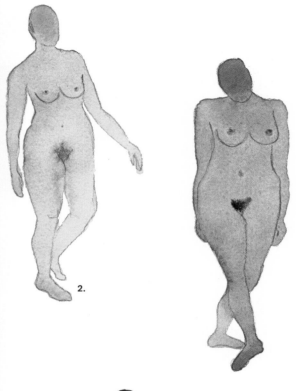

2.

3.

6.

A PALETTE OF COLOURS

In 1975 Thomas B. Fitzpatrick, a scientist at Harvard Medical School, came up with a standardised scale of human skin tones ranging from one to six. Obviously we have more skin colours on our planet than this, but Fitzpatrick's scale, however old-fashioned it may seem, is still used in dermatology. The scale is divided up into skin *phototypes*, based on how the skin reacts to sunlight.

People with skin types 5–6 have about sixty times more protection against DNA damage than those with skin types 1–2. So to have the same protection against sunlight, light-skinned people need to apply factor 50 sunscreen at least.

THREE SMART LAYERS

SKIN PHOTOTYPES

1. Very light skin tone: often freckled; burns easily and doesn't tan.
2. Light skin tone: usually burns, tans minimally.
3. Beige skin tone: sometimes burns, but also tans easily.
4. Light brown skin tone: rarely burns and always tans.
5. Dark brown pigment: hardly ever burns.
6. Black pigment: never burns.

The Dermis

In contrast to the thin epidermis, the dermis is up to ten millimetres thick. It's at its thickest on our backs. The dermis is the second layer of our skin, performing a range of specific functions. It maintains skin stability, controls our body temperature, supplies the epidermis with oxygen and nutrients, conducts signals from the touch cells (Merkel cells) to the brain, and plays a significant role in the immune system.

COLLAGEN, ELASTIN, AND HYALURONIC ACID

The two components of the dermis are collagen and elastin. These are proteins that give the skin its suppleness and elasticity — the keys to a healthy, youthful-looking skin. The collagen and elastin in the skin are surrounded by a jelly-like substance containing hyaluronic acid. With its unique humectant properties, hyaluronic acid helps maintain skin volume. The cells that produce elastin, collagen, and hyaluronic acid are called fibroblasts. As we age, they produce less of these substances. Fibroblasts are also affected by external factors, such as the sun, that trigger the enzyme matrix metallo-proteinase (MMP), which breaks down the strands of collagen and elastin in the skin.

This is why our skin becomes looser and more wrinkled as the years pass. The effect is most visible in the thin skin under our eyes, where the sun's UVA rays penetrate more deeply. This is the area where most people discover their first wrinkles.

SEBACEOUS GLANDS AND HAIR FOLLICLES

The sebaceous glands and hair roots are embedded in the dermis. Each sebaceous gland is connected to a hair follicle and supplies the epidermis with sebum, which combines with salts from sweat and acid from bacteria to form the hydrolipidic film.

Over-stimulated sebaceous glands that produce more sebum than the pores can handle are often associated with problems like acne and oily skin. The face and the head as a whole have a higher proportion of sebaceous glands than the rest of the body, with a particularly high density in the very visible T-zone comprising the forehead, the nose, and the chin.

Hair roots play a more important role in the skin than you might imagine. The follicles that hold them cohabit with the sebaceous glands. Hair follicles also contain the stem cells of melanocytes, the source of skin pigmentation. These cells migrate from the hair follicles to the epidermis, occupying its deepest layer (the basal layer). They can secrete pigment, especially if the skin is exposed to sunlight.

SWEAT GLANDS AND BLOOD VESSELS

The body's three million sweat glands are situated in the dermis, in almost all parts of the body. Exceptions include the lips, ear canal, and the glans. Although sweat glands can secrete as much as several litres of sweat if need be, the normal amount is 100–200 millilitres a day.

The dermis also contains an extensive network of blood vessels to supply the whole skin with oxygen and nutrients. The blood vessels are complemented by the lymphatic vessels. These collect liquid pumped out of the tissues and transport immune cells to keep everything running smoothly. If you've cut yourself and start to bleed, that means you've got through to the dermis; there are no blood vessels in the epidermis.

FIBROBLASTS

The cells that produce elastin, collagen, and hyaluronic acid are known as fibroblasts. The older we get, the less they produce. The illustration above shows what fibroblasts (in red) look like under the microscope.

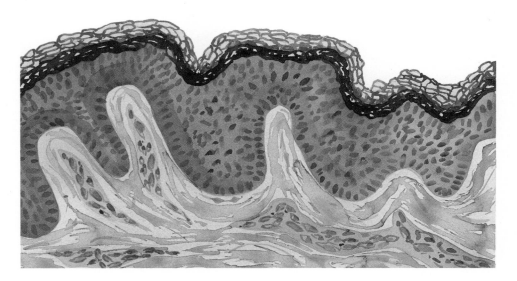

CROSS-SECTION OF SKIN TISSUE

If a biopsy is needed for some reason, colouring agents are added, as shown above. The illustration shows the epidermis in mauve and the cell nuclei in purple (keratinocytes, melanocytes, and fibroblasts). The reddish-purple colour shows the stratum corneum on the surface, while the area in pink is the collagen-rich dermal layer.

Hypodermis — The Deep Layer

The hypodermis is the part of the skin often referred to as subcutaneous fat. It's the deepest of the three skin layers, designed to insulate us against the cold. The cells that store fat are known as adipocytes. The hypodermis also contains a large number of vital blood vessels, nerves, and lymph vessels. The blood vessels run from the hypodermis to the dermis, while the nerves run even further, from the hypodermis to the epidermis. Some nerve endings are free, while others are connected to touch cells (Merkel cells). We have different types of sensory receptors for different types of touch and sense perceptions — light touch, vibration, pressure, and stretching — and free nerve endings that sense pain.

It's striking how the hypodermis varies in thickness from one part of the body to another. During my research, I've dissected many types of skin removed in the course of plastic surgery. The hypodermis covering the lower abdomen is often as much as three centimetres thick, yet there's hardly any in the face, particularly the eyelids. A lot of research is being done to see whether fat production in the face can be boosted as an antidote to loose, sagging skin.

CELLULITE

In contrast, there are other areas where people are keen to *reduce* the thickness of the hypodermis. The deepest layer of skin is home to cellulite. Eighty-five per cent of all women have cellulite, which is far more common in women than in men. Oestrogen is thought to be a strong contributing factor. Moreover, men have thicker skin, which conceals any cellulite they may have.

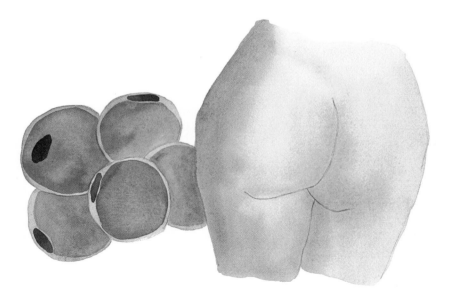

ADIPOCYTES
The cells that store fat (adipocytes) are globular in shape.

The structure of hypodermic tissue also differs in men and women. Men's hypodermic tissue displays a horizontal crosslinked pattern that squeezes fat together, whereas the connective tissue in women's skin has a honeycomb pattern, making any cellulite more visible.

Cellulite is completely natural, and it's there for a good reason — if a woman falls pregnant, her body needs to be able to form a reserve layer of fat. This also accounts for the skin structure that is typical of women.

Why Does Skin Scar?

Our skin protects our bodies. In turn, it's protected by a psychological mechanism that inhibits us from pricking holes in it. If you've ever had an injection or given blood, or if you recall having accidentally nicked yourself with a kitchen knife, you'll know just what I mean.

If an accident does occur, however, a fascinating process of healing begins that falls into three phases: inflammation, regeneration, and restructuring. We can learn a lot about the three smart skin layers by studying how wounds heal.

The inflammation phase begins as soon as the tissue is damaged and we start to bleed. The platelets, or thrombocytes, in our blood have a dual role: to stop the bleeding and to recruit immune cells, which play a central role in healing wounds. These cells cleanse the wound of bacteria, dead tissue, and foreign substances.

At the same time, the dead cells give off substances that irritate free nerve endings, causing pain. The skin produces various proteins, swiftly forming a temporary seal over the wound, and the inflammation continues until the damaged area has been cleansed.

In the regeneration phase, skin cells start to multiply in the area around the wound. Regeneration of skin cells results in new fibroblasts and keratinocytes. The fibroblasts produce collagen, thereby renewing the deeper layers of the dermis. The epidermis is regenerated by new keratinocytes. During the regeneration phase, the wound also begins to close gradually. This part of the process can take anything from a few days to several weeks.

This is followed by the restructuring phase, involving long-term skin repair through the restoration of the mix of collagen, elastin, and hyaluronic acid in the dermis.

Scarring results from the bundles of collagen that form to stabilise the wound as much as possible.

THE HEALING POWER OF SALIVA

Sores and ulcers in the mouth heal much faster and with less visible scarring. One of the reasons for this is the presence of saliva. Saliva contains special proteins and peptides that kill bacteria and speed up cell movement, thus making lesions heal faster. This is why animals often lick their wounds — and we have just the same reflex if we cut a finger, for example. Many research projects are under way to synthesise the various substances in saliva, so as to improve treatment to heal wounds, sores, and cuts.

THREE SMART LAYERS

Get to Know Your Skin Status

What we mean by 'status' is the appearance, quality, and visible characteristics of your skin. But skin status also depends on your subjective perceptions of your own skin.

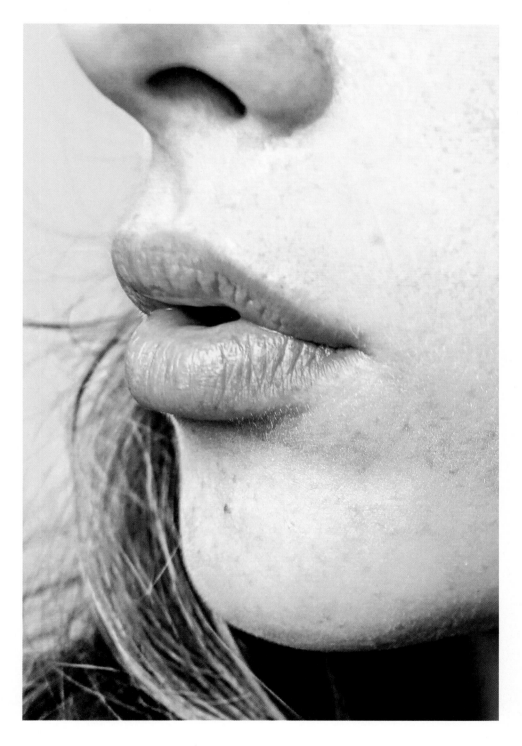

GET TO KNOW YOUR SKIN STATUS

Normal Skin

Normal skin is resilient, soft, and pliant. People who view their skin as normal have no obvious problems with dry or greasy areas of the face. Their skin is balanced. However, if you ask women today whether they think they have normal skin, most will probably answer no.

Sensitive Skin

Since there's no clear definition of sensitive skin, it's considered a subjective issue. Yet at least half of us experience skin sensitivity. Research outcomes vary from one part of the world to another. In 2013, dermatologist Howard Maibach conducted a wide-ranging review of studies of sensitive skin, arriving at a surprising conclusion: no less than 70 per cent of the world's population claim to have it. Some studies also show a gender gap, with sensitivity being significantly less common among men. Typical symptoms are itching, burning, or stinging, sensations that recur constantly. Those who rate their skin as sensitive most often report strong reactions in contact with cosmetics, skin creams, soap (mostly containing substances that aren't usually classed as irritants), and sunlight. People with sensitive skin often report that the condition worsens during dry, cold spells.

TRPV1

New studies suggest that there are complex reasons for perceived skin sensitivity. Particularly sensitive people may experience skin irritation because their epidermis is thinner, especially if they're exposed to water-soluble chemicals. It's also been noticed that such people show changes in the TRPV1 receptor (transient receptor potential vanilloid 1), which is located on nerve endings in the skin and in keratinocytes.

The best-known triggers of TRPV1 are high temperatures and capsaicin, the substance in chili pepper that burns our tongues. People with sensitive skin have higher sensitivity to TRPV1, and there are good reasons to suspect that several substances in skincare products have a triggering effect. A new study has shown that we become less sensitive and experience fewer skin irritations after the age of fifty. But if you don't want to wait until later in life, you can take steps to relieve sensitive skin by limiting your use of skincare products to a minimum (to avoid trigger substances as far as possible) and making use of substances that inhibit the TRPV1 receptor.

Dry Skin

This is particularly common in Scandinavia, and more common in the northern hemisphere in general, due to the colder, drier climate. Dry skin feels tight and easily irritated, and it looks chapped in places. The secretion of certain oils and humectants in the hydro-lipidic film is defective in this type of skin, so moisture evaporates more rapidly. Unlike normal skin, which is covered by an emollient film of naturally occurring lipids that prevent evaporation, dry skin needs extra help with these functions.

Oily Skin

Oily skin is a very common skin status which can affect both sexes from puberty up to the age of sixty. The greasy texture is the result of excessive sebum production. This type of skin looks oily and shiny, and often has enlarged pores. When consumer studies are carried out

on people with this skin type, they often say they feel uncomfortable in their own skins, even unattractive.

Oily skin can sometimes produce whiteheads, blackheads, or spots. The sebaceous glands are triggered mainly by testosterone, a male hormone occurring in both sexes. Oestrogen has been shown to reduce sebum production and the activity of the sebaceous glands, especially if taken in high doses. This is one possible reason why oily skin and acne can often be relieved by taking the contraceptive pill, which contains oestrogen.

Sebum production also varies in the course of the menstrual cycle. In one study of women with oily skin, sebum production was higher both in the week before and during the week of menstruation. The women experienced their lowest levels of sebum production in the second week of their cycle. However, no significant changes in sebum production were measured during the menstrual cycle of women with normal or dry skin. The protective layer of sebum that characterises oily skin can also be good for the skin. Interestingly, there's a link between oily skin, fewer wrinkles, and better-hydrated skin.

Combination Skin

Combination skin is so called because it combines two skin types: normal to dry skin on the one hand, and oily skin on the other. The oily skin is often shiny, particularly the T-zone comprising the forehead, nose, and chin. The skin of the cheeks, though, is generally normal or dry.

OILY SKIN = FEWER WRINKLES

Research shows that a relatively high level of oil in the skin is associated with fewer wrinkles. So don't worry about your greasy, shiny skin. It may well turn out to be the next beauty ideal!

Medical Complaints

All of us, I think, can identify with the nasty surprise of finding a pimple on our face just before a party. People who're constantly afflicted by pimples (acne) can suffer from social anxiety and experience their condition as a stigma. The psychological burden of medical conditions like acne, rosacea, vitiligo, psoriasis, and eczema is often so hard to bear that sufferers want to hide their skin at all costs — by applying concealer, for example.

Acne

Pimples originate in the sebaceous glands that lie next to hair follicles. It's normal to have acne as a teenager, when there are high levels of hormones in the system.

Studies show that pimples have more of an impact on women than on men. Female sufferers say their contacts with friends suffer, and they experience less interest in their studies, difficulties in forming romantic relationships, and a lack of intimacy in their relationships overall.

You might think not very many people have such serious symptoms, but that's not true. In fact, there's been a huge increase in acne worldwide. Today, 80 per cent of all adolescents and up to 65 per cent of all adults suffer from this inflammatory skin condition at some point. In the United States, acne is the most frequent reason for consulting a doctor.

In many parts of the world where acne was once a marginal problem, or even non-existent, it's now on the increase. There are many possible reasons for this rise, and I'll be looking at several factors in this book: our food, the use of various skincare products and make-up, our lifestyle, and a microbiome that's out of balance. It's interesting to note that acne is more of a problem in precisely those countries where skin cleansing and cosmetic use are widespread. The idea that cleansing is a way to tackle acne is nothing but a persistent myth. Western dietary habits are another major contributing factor.

There are a range of anti-acne treatments available today. The strongest treatment — which has a number of side-effects — is a high dose of synthetic vitamin A in cream, liquid, or gel form. The name on the pack-aging is tretinoin, and a doctor's prescription is required. Acne is also treated with antibiotics in many cases. In some cases, doctors may also prescribe certain contraceptives to reduce the activity of the sebaceous glands.

Rosacea

Rosacea is a common chronic inflammatory skin disorder that affects mainly adults. There are four types. The most common one is characterised by periodic flushes and lasting marks (erythema) in the central area of the face. Over time, this chronic flushing results in the formation of external blood vessels. The second type of rosacea involves permanent reddening of the skin, with sporadic pimples. The third type involves enlargement of the nose, with enlarged pores. The fourth results in inflammation of the outer parts of the eye, with symptoms such as a dry sensation, reddening around the eyelashes, increased tear production, and styes.

Rosacea affects both men and women, and over 5 per cent of the world's population experience it in one of the forms described, mainly between the ages of forty-five and sixty.

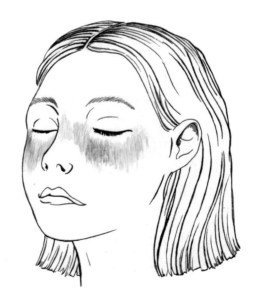

In a global perspective, it's over-represented among people living in northern Europe and people of Celtic origin.

A condition related to rosacea is perioral dermatitis, which manifests as reddened skin, small bumps, and a pimply rash around the mouth and sometimes under the nose.

The exact reason for rosacea remains unknown. While heredity appears to play a role, the condition has also been thought to be linked with disorders of the immune system and heightened nerve sensitivity.

Psychological stress is also a contributing factor. In fact, dermatologists in Sweden reported more cases of rosacea during the Covid-19 pandemic. The reason for this increase is believed to be higher levels of psychological stress.

There are indications that rosacea sufferers have higher levels of *Demodex* mites living in their skin. We always have some mites (a kind of microscopic arachnid) in our skin. They thrive both in bed linen and skin. *Demodex* mites live in our hair follicles and sebaceous glands. According to one theory, their faeces trigger an inflammatory response in our skin.

There's also a medical link between rosacea and autoimmune diseases such as diabetes, coeliac disease (gluten intolerance), and rheumatism.

Though rosacea hasn't been studied anywhere near as much as acne, new research shows that its psychosocial impact can be very serious, leading to social anxiety and depression. Interestingly, men tend to experience worse psychological effects than women.

Today, the condition is often treated with antibiotics and azelaic acid, which both have anti-inflammatory properties. The drug brimonidine is used to alleviate skin redness. This treatment reduces the blood supply to the skin, preventing it from reddening so much. Ivermectin, an insecticide that kills *Demodex* mites, is one of the newer treatments for rosacea. It's not known exactly why antibiotics affect rosacea, but one possible explanation, if the theory is correct, is that mites' faeces enable bacteria to grow in the skin.

Psoriasis

The symptoms of psoriasis can vary. In its most common form, plaque psoriasis, which accounts for 80 per cent of all cases, the symptoms are well-defined red and scaly rashes found mostly on the elbows, knees, and scalp. The scalp is affected particularly often.

Can anyone get psoriasis? The answer is no: it looks as if the condition is genetic in origin and often linked with autoimmunity. In other words, the body's own autoimmune system is out of kilter. By attacking the body's

MEDICAL COMPLAINTS

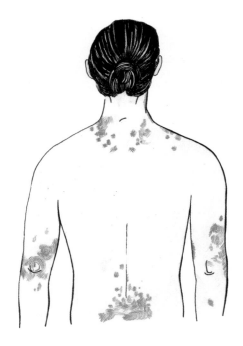

own cells, it brings on psoriasis.

Psoriasis can begin at any time of life, but in most cases, it's thought to start before the age of twenty-five. There's no apparent difference between men and women. The disorder affects 3–4 per cent of the population in Europe. It's uncommon in children, but can be found in at least 5 per cent of the over-fifties. Geographically speaking, its prevalence declines from north to south, with fewer cases in southern Europe and in the southern hemisphere. It's often said the figures for Asia lie between 0.1 and 0.5 per cent of the population. But these statistics are unreliable, as less serious forms of the disease often remain undiagnosed. Psoriasis can be hard to live with; this is partly because it affects sufferers' daily lives (psoriasis plaques can be itchy), but also because the hard, reddish plaques of skin can be very noticeable, affecting sufferers' appearance. Studies show that psoriasis patients are more prone to depression than the rest of us.

The most common antidotes to psoriasis today are creams containing vitamin D analogues, such as calcipotriol, which block the growth of keratinocytes and inhibit plaque formation. Cortisone creams to reduce skin inflammation are also in widespread use. Light treatment (UVB) is also used to reduce inflammation and inhibit keratinocyte growth. In severe cases, doctors may consider treatment using immunoregulatory drugs (methotrexate or biological drugs).

A bacterial imbalance between gut and skin is a possible factor in psoriasis. I'll touch on this in the chapter about the skin microbiome.

Eczema — Atopic Dermatitis

Eczema is a chronic inflammatory condition that manifests as dry and itchy areas in different parts of the body. It's the most common inflammatory skin disorder, affecting 15–30 per cent of all children and 2–10 per cent of all adults. Eczema is connected with a mutation in one of the skin's proteins, filaggrin, which can also show up as allergies or asthma. In addition, there's a clear link with our Western lifestyle.

Eczema also appears to involve a gut–skin connection, which we shall describe in more detail in the chapter on the microbiome and lifestyle.

The most common treatments for eczema today are emollient creams and humectants, and various cortisone creams or tacrolimus, a new immunoregulatory substance. Eczema is also treated with sunlight. Any eczema sufferer who has travelled to a sunnier part of the world will know what relief sunlight can bring.

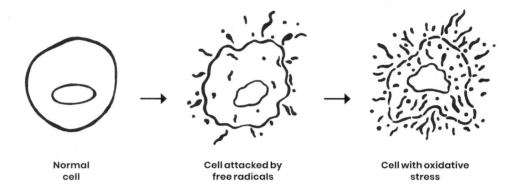

**Normal
cell**

**Cell attacked by
free radicals**

**Cell with oxidative
stress**

MEDICAL COMPLAINTS

Oxidative Stress

Recently, expressions like oxidative stress, free radicals, and antioxidants have become common currency. But what do they actually mean?

Oxidative stress is a natural process and part of cellular respiration. We need oxygen to live. When we inhale it, the body forms particularly active molecules called free radicals. Free radicals are prone to react with everything around them, and if there are too many they damage cell membranes and the DNA in the cell's nucleus. These harmful molecules are neutralised by antioxidants.

We have many antioxidant enzymes in our skin, three of the most important being superoxide dismutase (SOD), catalase, and glutathione peroxidase (GPX). For these antioxidant enzymes to function properly, we need minerals such as manganese, zinc, selenium, and copper. The skin also contains other substances which can neutralise free radicals, such as vitamin C and vitamin E.

Many skin complaints, as well as skin ageing, are linked with oxidative stress. In certain disorders, oxidative stress destroys pigment cells. In vitiligo, for example, pigment disappears in patches, resulting in white blotches on the skin. Natural oxidative stress can be aggravated by external factors such as sunshine, pollution, or other toxins.

ABOUT THE SKIN

Vitiligo

If melanocytes are attacked by the body's own immune system, in combination with high oxidative stress (see opposite page for definition), they lose their dendrites (arms), whose main purpose is to disseminate pigment evenly throughout the skin. The resultant condition, vitiligo, manifests as white patches of skin. It's a fairly common disorder, affecting 0.5–2 per cent of the population. However, there may be a lot more unreported cases, as many people don't seek medical care for changes in skin pigmentation.

Vitiligo patches have a higher concentration of oxidative stress, owing to the presence of hydrogen peroxide and other free radicals. The German dermatologist and skin scientist Professor Karin Schallreuter MD is a specialist in vitiligo.

Winnie Harlow, who has made vitiligo her trademark, is one of today's most sought-after photographic models.

Since vitiligo isn't itchy or painful, and the main reason to seek treatment is that it affects people's appearance, Winnie is a great role model who has transformed an imperfection into a unique personal strength.

There's no standard treatment for vitiligo yet. However, the next page touches on treatments involving PC-KUS cream and climatotherapy. Before starting to investigate vitiligo, I tried various treatments which are still being tried out on patients — tacrolimus ointment, light therapy, and even having melanocytes transplanted from areas of skin with normal pigment — but none of them worked. In fact, my skin got worse.

Melasma — Vitiligo in Reverse

A drastic increase in the pigmentation of sizeable areas of the face usually indicates melasma. Some women develop large brown patches on their faces during pregnancy or as a result of taking the contraceptive pill in summer. This is because melanocytes are hormone-sensitive. Higher levels of female hormones in combination with sunlight produce irregular pigmentation. The only thing that helps is to use a very high-factor sunscreen or to stop taking hormonal contraceptives. The condition may also clear up after giving birth if it progressed during the pregnancy.

In some cases, melasma is linked with a relatively inactive thyroid gland. This is hard to treat, and sadly there's no miracle cure. Dermatologists prescribe azelaic acid or isotretinoin in ointment form, but the results are variable. Unfortunately, many products available online can have really harmful effects. One of the substances they contain is hydroquinone, which can actually be counter-productive, producing blue-black pigmentation, or permanent white patches resulting from the destruction of melanocytes.

In short, beware of buying anti-melasma creams online!

Winnie Harlow, who has made vitiligo her trademark, is one of today's
most sought-after photographic models.

Johanna's Experience of PC-KUS Treatment

I've had vitiligo since my childhood. I visited numerous dermatologists and tried everything from cortisone, tacrolimus, and UVB treatment to transplantation of healthy pigmented skin cells to the affected areas. Nothing worked, and the vitiligo actually got worse from all the treatments.

I studied in Uppsala to become a pharmacist and graduated in 2002. During my years at the university, I got in contact with Karin Schallreuter and this gave me the opportunity to conduct doctoral studies in clinical and experimental dermatology with Schallreuter's team.

Based on the discovery of the 'butterfly bioterins' in the vitiligo spots, Professor Schallreuter and her team developed a treatment to reduce oxidative stress in skin. When the treatment was used, the melanocytes faded and began producing pigments again. The treatment did wonders for my face with 90% repigmentation.

The hands and feet, however, are still difficult areas to treat. Despite its uneven results, however, Karin's therapy remains unique in actively tackling oxidative stress in the skin and helping to even out patches.

This treatment was developed in 1995 and has since helped thousands of patients at three different locations: Medical Clinic — The Dead Sea, Institute for Pigmentation Disorders by V. Greifswald, Germany, and at the University of Bradford, England. In 2006, I defended my thesis on vitiligo and received a doctorate in experimental dermatology. Today I feel happy about my vitiligo and enjoy the white spots, even though several of them have now almost completely disappeared. If I'd not had them, I probably wouldn't have the levels of motivation I do, or met all the amazing people and researchers that I have over the years, or had so many wonderful and varied experiences.

Professor Karin Schallreuter Prof, MD — Mother of Vitiligo

Professor Karin Schallreuter has studied vitiligo for the past 25 years and has been a working dermatologist and specialist at the Mayo Clinic, Minnesota. She is one of the founders of the Center for Skin Science at the University of Bradford, as well as the founder of the Institute for Pigmentary Disorders in Greifswald, Germany. In the 1980s, when Professor Schallreuter and her late husband, ProfessorJohn Wood, decided to investigate the causes of the white patches that characterise vitiligo, they soon found a link with unusually high levels of oxidative stress. Examining melanocytes under the microscope, they could see that the affected cells no longer had any dendrites (arms). As you probably recall, melanocytes are the cells that secrete pigment in the skin, while dendrites are responsible for spreading it evenly.

Professor Schallreuter and a team of American and German scientists discovered that specific substances are produced in the white skin. These substances emitted fluorescence when exposed to light from Professor Wood's lamp. They belong to the biopterins family and are also what give some butterflies their fluorescent color.

The discovery of the ability for vitiligo to fluoresce was published in the prestigious journal Science in 1994.

Professor Schallreuter has authored more than 170 scientific articles in experimental and clinical dermatology and treated more than 6,000 patients with vitiligo worldwide. Together with Professor Wood, she invented Pseudocatalase (PC-KUS) — a topical treatment with a focus on mimicking the natural antioxidant enzyme in human skin, catalase.

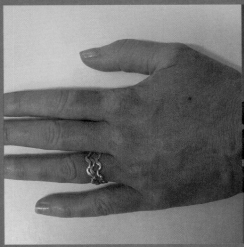 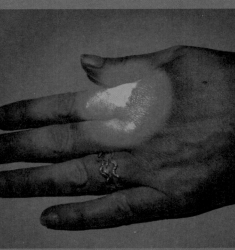

Vitiligo on hand in visual light.

The same hand under Wood's lamp gives a white fluorescent light.

MEDICAL COMPLAINTS

What You Didn't Know About Vitiligo

There are three different types of vitiligo:

- Vitiligo vulgaris (generalised vitiligo) is the most common form, and is characterised by white spots that appear symmetrically on both sides of the body.
- The second type, vitiligo segmentalis (segmental vitiligo), appears as spots only on one side of the body.
- The third type is called vitiligo focalis. It is characterised by small patches located on one or a few parts of the body.

Researchers are conflicted about the reason vitiligo occurs. Some argue that it is caused by autoimmunity, i.e. when your immune system attacks the pigment cells. Others claim that a high concentration of oxidative stress destroys pigment cells. Most likely, it is a combination of both. Patients with vitiligo (especially with vitiligo vulgaris) have antibodies against the pigment cells as well as a high concentration of hydrogen peroxide, which is an oxidative marker. What came first, the chicken or the egg? We do not know.

For a long time, the assumption was that the white spots didn't contain any melanocytes at all, but research has shown that melanocytes are still present: they've lost their ability to form pigments, though.

Because depigmentation can occur in visible skin areas, the disease can be socially stigmatising and cause psychosocial concerns, with, among other things, impaired well-being.

Treatment of vitiligo

No uniform national guidelines for the treatment of vitiligo exist today. Most often, no further medical treatment is offered, but the focus is on patient information and information on disease progression.

However, some dermatologists prescribe cortisone or tacrolimus, a substance developed for treating atopic dermatitis, which you are instructed to apply to the affected areas. Evaluation takes place after six months. A few patients see some effect due to the immunomodulatory effect. A light treatment with UVB light is prescribed in some regional areas, with varied effects.

What to avoid if you have vitiligo

- Swimming in pool water containing chlorine — this can aggravate vitiligo.
- Skincare containing the active ingredient Q10 (INCIi: Ubiquinone). Studies show that applying creams with Q10 can aggravate vitiligo.
- Stress — this is easier said than done, but it's been shown that vitiligo is greatly aggravated by mental stress.
- Avoid getting injured. People with vitiligo are especially vulnerable to injuries and pressure on the skin, which often causes a new white spot where the wound or scab

was. This is a well-known phenomenon, called the Köbner phenomenon.
- Avoid tight clothing and tight shoes. Vitiligo often manifests itself around the waist, under the bra, and on the feet, due to constant pressure on the skin, which can trigger pigment loss.
- Avoid smoking. Smoking has often been shown to aggravate vitiligo, especially around the mouth. It is also difficult to treat vitiligo if the patient smokes, since the repigmentation happens very slowly in these cases.

Do not buy Pseudocatalase online!

To date, Professor Schallreuter has treated 6000 patients with vitiligo with NB-activated pseudocatalase. Several cosmetics companies sell creams online named pseudocatalse. Unfortunately, this is not the same composition as the original cream, and therefore does not give the desired effect.

MEDICAL COMPLAINTS

ABOUT THE SKIN

Johanna Visits the Dead Sea ...

At Last — A Sea Change in Vitiligo Research

The Dead Sea, which lies between Israel, the West Bank, and Jordan, is an ideal location for dermatological research of various kinds, including studies of psoriasis and vitiligo. Its position 400 metres below sea level affects the way in which the sun's rays fall on the region.

What makes the Dead Sea region so different is the ratio of UVA to UVB light. It receives more UVB than UVA, and the UVB rays have a particular wavelength that triggers melanocytes and keratinocytes, with a beneficial impact on both vitiligo and psoriasis. This treatment is generally referred to as 'Climatotherapy'.

Inspired by the Dead Sea area, Karin Schallreuter invited several hundred people with vitiligo a year from all over the world to participate in her research. For several years, I took part both as a scientist, which proved incredibly interesting. It was a wonderful time. The days began and ended with a dip in the Dead Sea, while I spent the hours in between isolating blood cells, measuring pH values, and analysing oxidative stress in the skin of several hundred patients.

The treatment consisted of bathing (or, more accurately, floating) in the Dead Sea for fifteen minutes twice a day; applying PC-KUS (more on p. 35), and exposing the skin to moderate UV light early in the morning and late afternoon for 15 minutes each side. For twenty days, the Dead Sea was my home, as well as the home of many patients. In addition to a strong pigmentary response, our studies showed that these treatments were also an effective tool for long-lasting improvement in quality of life and patients' well-being.

Even at day twenty, quality of life was significantly improved compared with day one, and this effect was still significant after twelve months. Moreover, social anxiety, anxious-depressive mood, and helplessness — which are common features for patients with vitiligo — were all significantly reduced.

MEDICAL COMPLAINTS

Senescence — When the Skin Ages

While you can have firm and perfectly hydrated skin at twenty, expecting your facial skin to have the same elasticity and your forehead to be equally smooth at forty is unrealistic.

The skin ages at the same rate as our other organs, but differs in that it's constantly exposed to the elements. However, there's a lot we can do to keep our skin in better shape and make sure it doesn't age faster than the natural rate.

As we age, the skin's cells gradually enter a new stage known as senescence. Cellular senescence means the cells cease to function as normal. They stop dividing, become dysfunctional and passive, and can no longer absorb enough nutrients.

Once the process is under way, it usually advances rapidly. Its onward march will be particularly noticeable to anyone who's already found their first grey hair. Greying begins with one strand of hair and spreads in a surprisingly short time.

Research is currently under way to find out whether senescent cells can be removed, which would be the ultimate anti-ageing treatment. This technique is known as senolytics (see p. 159).

Inflammaging

As we age, the body's ability to produce antioxidants declines, resulting in higher oxidative stress in the skin. Increased oxidative stress sparks off the ageing process, so that more and more cells end up in a senescent, passive state.

There's a two-pronged effect: natural ageing inevitably pushes cells into the senescent stage, and the process is accelerated by external factors like sunlight and pollution. The increase in the number of senescent cells causes chronic low-intensity inflammation in the body, which scientists call 'inflammaging'. The term refers to inflammation as a result of oxidative stress, which leads to ageing in combination with natural senescence.

Now let's take a closer look at these effects in the light of intrinsic ageing (ageing from within) and extrinsic ageing (which is caused by external factors).

Intrinsic Ageing

If we weren't affected by any external factors, the skin would age significantly more slowly than it does under real-life conditions. Natural skin ageing, which begins quite late in life, is characterised by smooth, pale, drier, and less elastic skin with fine wrinkles. Intrinsic ageing is essentially hormonal.

HORMONES

When women reach menopause and men andropause, they experience an imbalance in their sex hormones. In women, this is shown by a dramatic change in skin elasticity, caused by a decline in the production of oestrogen, a hormone that stimulates collagen in the dermis. In men, the process is more gradual, and results from falling testosterone production. In most cases women don't reach the menopause until they're nearly fifty. However, many notice a significant change in their skin as soon as they hit forty.

Menopause, which arrives at around fifty, is preceded by a period between the ages of thirty and forty known as adrenopause. This transitional stage is characterised by a gradual decrease in dehydroepiandrosterone (DHEA), a hormone secreted by the adrenal cortex. DHEA, which is closely linked with ageing, is needed in the body to secrete both oestrogen and testosterone. We know its concentration is high at birth, after which it falls, rising again during puberty, and then falling between thirty and forty. Although there are corresponding changes in both sexes, the concentration of DHEA is lower in women. DHEA both stimulates collagen production and inhibits its breakdown.

By the time hormone levels in the skin finally fall, intrinsic ageing is under way. The decline in collagen levels causes wrinkles to form little by little. This normally happens between thirty and forty.

ANTI-AGE TREATMENT WITH DHEA

Is DHEA the ideal anti-ageing substance? Studies of the results of applying it to ageing skin have shown positive effects, but DHEA also stimulates the sebaceous glands, so it can lead to oily skin and unwanted hair growth. Supplements containing DHEA have also been linked to hormone-sensitive cancers such as prostate and breast cancer. Its use is therefore not recommended.

THINNER SKIN LEADS TO WRINKLES

The epidermis becomes thinner once the keratinocytes no longer reproduce at the same rate as before. This is noticeable at an earlier age in women than in men, especially in areas such as the face, the neck, the upper part of the chest, and the hands. Skin thickness decreases by about 6.4 per cent a decade. Thinner skin makes us more sensitive to sun, as sunlight can penetrate a thin skin more easily.

The dermis also becomes thinner with age. With fewer fibroblasts to produce collagen, elastin, and hyaluronic acid, the network formed by these substances is reduced, resulting in a less pliant skin. You can test this by pinching your skin and seeing how long it takes for it to bounce back to its original position. The longer it takes, the more advanced the ageing process.

Thinner skin also makes blood vessels more visible, and they can end up looking somewhat like a spider's web.

UNEVEN PIGMENTATION

Each decade after the age of thirty, the number of pigment-producing melanocytes in the epidermis falls by between 10 and 20 per cent, which leads to uneven pigmentation. In addition, melanocytes can also become over-active. One theory is that UV light makes them go slightly mad. Over-active melanocytes produce liver spots (*lentigo solaris*). As a result of these two processes, older skin has uneven pigmentation and gradually loses its glow.

DRIER SKIN

Sebum production can increase until people reach their fifties, after which it falls. The decline is significant (as much as 50–60 per cent), and results in a drier skin. As I'm sure you recall, sebum production is important in enabling the skin to build up an effective hydrolipidic structure to preserve moisture in the skin. This is why the skin's water content also declines with age.

LESS SWEAT

Ageing also goes hand in hand with a reduction in sweat gland activity and sweat production. In combination with a decline in blood flow to the skin, this can affect body temperature, which falls with increasing age.

Extrinsic Ageing

Some 80 per cent of all skin ageing is the result of exposing the skin to outside influences, particularly the sun. Features of skin subject to extrinsic ageing include deep wrinkles, coarse skin texture, a sallow complexion with patchy pigmentation, and loss of skin elasticity.

The first signs of extrinsic ageing can already be seen in fifteen-year-olds who've been exposed to too much UV light. In places where exposure to the sun is very limited, on the other hand, extrinsic ageing doesn't appear until around age thirty.

The sun accelerates the onset of

senescence in cells, resulting in faster breakdown of collagen and elastin. It's easy to see the results of sun exposure by comparing skin that's normally covered up (on the breasts or buttocks, for example) with areas such as the upper arms.

Smoking and air pollution are thought to come joint second as the main causes of extrinsic ageing after exposure to the sun.

UVA CAUSES WRINKLES AND UVB CAUSES IRREGULAR PIGMENTATION

Solar UVA radiation is seriously bad in terms of prematurely ageing your skin. These rays are so strong that they can even pass through a pane of glass. Google 'truck driver sun', and you'll see a face that's familiar to most people in the beauty industry. It belongs to a man who worked as a truck driver in California for twenty-eight years. His left side was exposed to

extremely high levels of UVA through the cab window, causing very serious sun damage. You can clearly see the sagging skin and wrinkles down one side of his face. He's sixty-nine in the picture, but looks at least ten years older. What this example also shows, however, is that the cab window kept out the UVB rays. If it had let them through as well, the man's face would have had many more discoloured patches. Unlike UVA, UVB reaches only the basal layer of the epidermis, where the melanocytes that secrete pigment are located.

Skin Cancer

There are several different types of skin cancer, but malignant melanomas (cancer of the melanocytes) are by far the most serious kind. Once melanoma cells move down into the dermis and spread into the lymphatic system, the resultant cancer is hard to treat. However, if the cancer hasn't spread and the lesion is removed locally, there's normally no big problem. That's why it's absolutely vital to have your skin checked and get moles identified early on. As soon as a mark on your skin changes in size or colour or starts acting differently (itching or bleeding, for instance), you need to see a dermatologist and get a diagnosis. There's a clear link between melanomas and the amount of sunburn you've had in the past. That link is particularly strong if you've had sunburn before the age of twenty-five.

There are around 2300 melanoma skin-cancer deaths in the UK every year (2017). In Australia, this number is approximately 1700, and in the US, 7230 (figures accurate as at 2019). Since women are usually diagnosed earlier than men, mortality is higher among male patients.

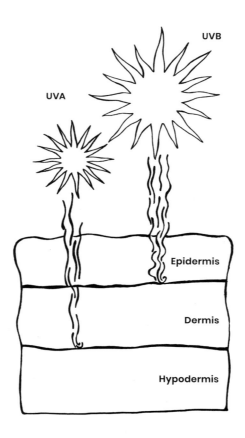

The UVA rays penetrate the dermis, while the UVB rays reach only as far as the basal layer between the epidermis and the dermis.

The reason for this difference is presumably that women pay more attention to their skin and seek treatment earlier. Remember: the earlier the diagnosis, the better the prognosis!

Another type of skin cancer is basal-cell carcinoma (BCC or basalioma), a cancer of the keratinocytes. This is less serious than malignant melanoma, and it hardly ever spreads. BCC is caused by exposure to the sun, and you often see it in people who've spent a lot of time outdoors in sunny weather.

If BCC is the kinder variant of keratin-ocyte cancer, squamous cell carcinoma (SCC) is the nastier variant. SCC is more aggressive than BCC, forms lesions at an earlier stage, grows and spreads to deeper tissues faster, and, like malignant melanoma, can spread through metastasis.

CAN THE SUN BE GOOD FOR US?

If we avoid getting sunburnt and stay in the shade, or limit our time outdoors to times when we're not exposed to the sun's most powerful rays, the sun is a wonderful thing. It's our best source of vitamin D. However, only a very short time in the sun is needed for sufficient vitamin D production. For example, white Caucasians across the UK need nine minutes of daily sunlight at lunchtime from March to September to retain adequate vitamin D levels throughout the winter. This assumes that forearms and lower legs are exposed June–August. The sun can also be good for people suffering from inflammatory diseases or autoimmune disorders. As we've already seen in this chapter, UVB light relieves autoimmune skin conditions such as psoriasis and vitiligo.

Scientists have also discovered that UV rays that touch the skin or the eyes send signals to the brain, setting off hormonal processes. The fact that many people feel better during the six warmest months of the year was once attributed solely to higher vitamin D production. New studies, however, show another possible reason: direct signals to the brain, which responds by producing feel-good hormones like serotonin. Again, though, it's important to stress — more is not always better.

Less is more!

TIPS FROM DR TORBORG

Dr Torborg Sturesdotter Hoppe is a dermatology specialist. Besides lecturing in dermatology and venereology at Uppsala University, she works at Stockholm's Skin Diagnosis Centre, where she sees patients with various kinds of tumours every day. This is what you need to look out for when you're checking for marks or patches on your skin, she says:

- Moles that spread and grow asymmetrically, develop projections, or become uneven in colour (more than two colours)
- New marks that are different from the ones you already have
- Sores or wounds that fail to heal, which could also be some kind of skin cancer — act if you find one!

SENESCENCE — WHEN THE SKIN AGES

About Skincare

The skincare sector is creative. It's built an industry worth billions using thousands of variants of the simple ingredients that go into a cream or ointment — water, oil, and additives. Many products do exactly what it says on the tin, but others miss the most important point of all — what's best for the skin.

What's in a Skin Cream?

Having removed her make-up, forty-five-year-old Helen carefully pats her face dry with a towel. She opens the bathroom cabinet, which is full of tubes, jars, and capsules. Taking out a little grey plastic bottle, she presses the attached pump a couple of times, then applies a silvery cream to her face. It's a night serum which supposedly 'helps the skin regain its elasticity'.

Then it's time for her eye cream.

Helen opens a tiny tube and carefully applies its contents to the area around her eyes, 'to reduce bags under the eyes and boost collagen production in the skin, minimising deep wrinkles'.

Now it's time for her night cream. She unscrews the plastic lid. The yellowish cream has a pungent scent. The label on the glass jar says it contains vitamin A and hyaluronic acid, a substance that 'hydrates the skin, giving it a youthful glow'. It also describes the product as 'paraben-free'.

But what exactly has Helen just put on her face? If we're to believe the sales pitch, it contains various vitamins, antioxidants, natural oils, and other substances that, it's claimed, will make her skin both smoother and firmer.

There are skincare products that can genuinely produce these results, and there are international rules that require skincare products to live up to their claims. Yet even so, many such products fail to achieve the desired result: additives meant to have a particular effect are present in too low a concentration to be effective, and some products contain skin irritants.

In fact, many companies focus mainly on profit maximisation by ensuring their products can be stored for a long time (which works best with stable products that can survive logistics and distribution), using cheap ingredients, and charging high retail prices. Although skincare is their business, they don't put the skin's best interests first.

In this chapter we'll lift the lid on today's skincare products and take a look at what they actually contain. We'll identify ingredients that are really effective — and others you should avoid if you want to have a healthy skin and keep it in good condition.

What Are Skin Creams Made Of?

A skin cream is basically a blend of oil and water. To produce a light cream, you start with an oil-in-water emulsion. To make a richer cream or a salve, you reverse the proportions, producing a water-in-oil emulsion. So the ingredients are essentially the same, it's just the proportions that are different. To make sure the mixture of oil and water holds together over time and doesn't separate, you add an emulsifier. Emulsifiers bind together liquids that are partly hydrophilic (attracted to water) and partly hydrophobic (with an affinity for oil). Skin cream can be compared to mayonnaise, in which the yolk (or, to be precise, the lecithin in it) functions as an emulsifier, keeping the oil and the liquid together.

So the only ingredients you need to make a skin cream are oil, water, and an emulsifier. However, as we all know, there are plenty of other names on the list of ingredients.

ABOUT SKINCARE

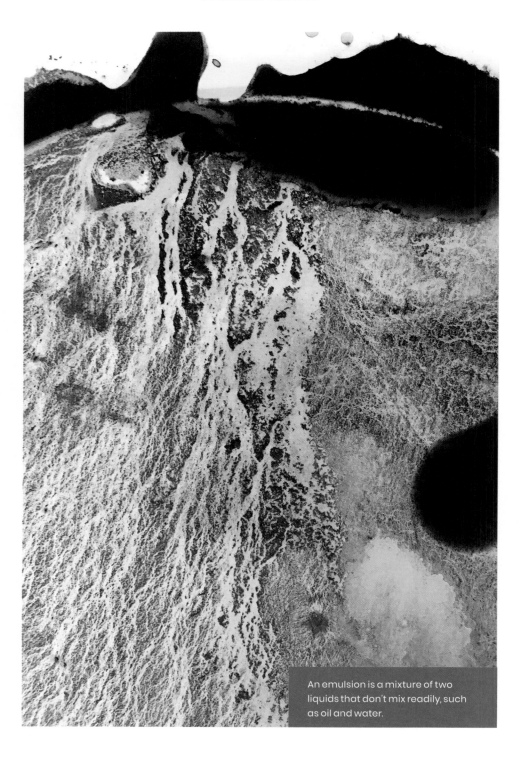

An emulsion is a mixture of two liquids that don't mix readily, such as oil and water.

What's the INCI List?

Learn How to Decipher the List of Ingredients

The ingredients of skincare products are listed on the back of the packaging, in what's known as the INCI list. The INCI list on a day cream has the same function as the list of ingredients on a loaf wrapper: it shows what's in the product.

Within the EU it's mandatory to use the INCI system to list the ingredients of all skincare and cosmetic products. The INCI (International Nomenclature of Cosmetic Ingredients) provides a set of standard terms, which are in general use in the UK, Australia, Canada, the US, Japan, and other countries too. Ingredients are listed with their standard names, in order of proportion, with those present in larger amounts listed first. This applies to all ingredients that account for more than 1 per cent of the product. Those accounting for 1 per cent or less are also listed, but they come at the end and not in any particular order. However, it's worth bearing in mind that even a substance accounting for only 0.1 per cent of the product can still affect the skin. One example is retinol, an active ingredient that has an effect even if it amounts to a mere 0.1 per cent of the product concerned.

The names of plant-based ingredients and the like are often Latin-based, while other substances are listed in English. INCI language is sometimes completely different from the common names we use for particular substances. That means that some natural ingredients can sound like chemical products, and vice-versa. INCI names don't indicate whether natural ingredients are organic. However, many manufacturers mark organic ingredients with an asterisk (*).

INCI terms are used in this book to give some examples of the most common substances. You'll find these listed just under the ingredients I describe.

WHAT'S IN A SKIN CREAM?

In This Chapter:

Skin Cream Ingredients Fall Into 7 Groups

Emollients, humectants, active ingredients, thickeners, preservatives, emulsifiers, and masking agents.

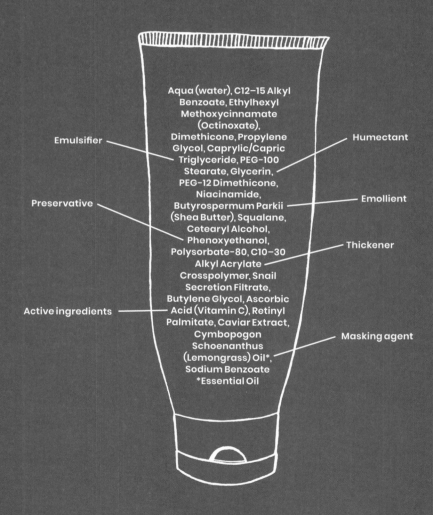

Emulsifier

Preservative

Active ingredients

Humectant

Emollient

Thickener

Masking agent

Aqua (water), C12–15 Alkyl Benzoate, Ethylhexyl Methoxycinnamate (Octinoxate), Dimethicone, Propylene Glycol, Caprylic/Capric Triglyceride, PEG-100 Stearate, Glycerin, PEG-12 Dimethicone, Niacinamide, Butyrospermum Parkii (Shea Butter), Squalane, Cetearyl Alcohol, Phenoxyethanol, Polysorbate-80, C10–30 Alkyl Acrylate Crosspolymer, Snail Secretion Filtrate, Butylene Glycol, Ascorbic Acid (Vitamin C), Retinyl Palmitate, Caviar Extract, Cymbopogon Schoenanthus (Lemongrass) Oil*, Sodium Benzoate *Essential Oil

Emollients

Skin cream contains emollients to preserve moisture in the skin, preventing it from becoming dry and flaky.

Emollients have an occlusive effect, which means they cover the top layer of skin and curb evaporation. The loss of water through the skin, which goes on all the time, is known as transepidermal water loss, or TEWL.

Normal skin retains moisture and remains soft thanks to the hydrolipidic film, a unique oily layer covering the outer-most layer of skin, the stratum corneum.

The sebum secreted by the sebaceous glands occupies the space between the cells, making the surface of the skin supple. This is the hydrolipidic film that maintains the skin's barrier function.

The emollient component of skin cream consists of natural oils or mineral oils like paraffin oil, waxes and butters, and its purpose is to imitate the hydrolipidic film as closely as possible, thereby contributing to a more effective barrier function in dry skin.

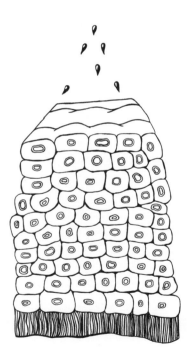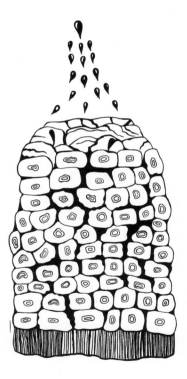

TEWL — TRANSEPIDERMAL WATER LOSS

Evaporation of moisture from healthy skin (on the left) and from damaged, dry skin (on the right).

EMOLLIENTS FOUND NATURALLY IN THE SKIN

A good tip is to look first and foremost for emollients that the skin itself produces and has an affinity with. Examples include ceramides, squalene, omega 3 and omega 6 fatty acids, and cholesterol. These are part of the sebaceous glands' beneficial blend of oils and water that, together with humectants, form the hydrolipidic film.

Several studies have shown that creams containing ceramides reduce skin inflammation, while also enhancing the barrier function. Natural ceramide levels in the skin change with the seasons. They're at their lowest in winter, which may be one reason that many people experience drier skin during cold spells. The same applies to omega 3 fatty acids and other lipids, which reach low levels during the coldest six months of the year. Creams containing ceramides and omega 6 have been shown to help psoriasis sufferers. Ceramide creams have also proven beneficial for children with eczema, improving hydration levels in the skin. As regards omega 6, people who are prone to spots have been shown to have lower levels of this oil in their skin.

The first article about applying cholesterol directly to the skin was published in 1983 in the prestigious scientific journal *The Lancet*. People with ichthyosis, a hereditary disorder characterised by extremely dry, scaly skin, saw an improvement in their condition after applying a cream with 10 per cent cholesterol. With its high cholesterol levels, lanolin, an oil obtained from wool, has been widely used in skincare. However, for ethical reasons the industry is now increasingly moving away from lanolin use.

Squalene, an oil that is naturally present in the skin (or a variant, squalane, which keeps better) is widely used in the skincare industry. It used to be extracted from shark's liver, but fortunately a way has now been found to obtain it from olives. Squalene is good for the skin and has very effective emollient properties.

INCI: Ceramide, Squalene, Squalane, Cholesterol, Linoleic Acid (Omega 6), Linolenic Acid (Omega 3)

VEGETABLE BUTTERS

What kind of vegetable butter you add to a skin cream depends on what kind of consistency you want. Shea butter has a low melting point and makes for a lighter cream, while cocoa (or cacao) butter, with its higher melting point, produces one that's more viscous. The advantage of using natural butters and oils is that they're often naturally rich in substances that are good for the skin, such as antioxidants and polyunsaturated fatty acids like omega 3. Shea butter, extracted from the kernels of shea nuts, is made up of triglycerides with oleic acid, stearic acid, omega 6, and palmitic fatty acids.

INCI: Butyrospermum Parkii (Shea) Butter, Mangifera Indica (Mango) Seed Butter, Theobroma Cacao Seed Butter, Cocos Nucifera (Coconut) Butter

WAXES

Waxes are widely used in creams for dry skin. In addition to their emollient and occlusive properties, they have a richer consistency.

INCI: Stearic Acid, Helianthus Annuus (Sunflower) Seed Wax, Simmondsia Chinensis (Jojoba) Seed Wax, Cera Alba (Beeswax), Lanolin

NATURAL OILS

The oils in your skin cream are either natural or synthetic or of petrochemical origin. The advantage of natural oils is that they contain valuable antioxidants and fatty acids. However, natural oils aren't always good for your skin — it's not that simple.

The nutrients found in these oils (in the form of polyunsaturated fatty acids) can sometimes result in allergic reactions in the skin when they turn rancid. Reactions vary from one individual to another. So natural oils can be good for some people, while others (usually those with sensitive skins) may react if the product is not handled correctly. Long-shelf-life product with weak/limited control on storage conditions may be better off formulated with more stable oils, such as silicons and mineral oils..

INCI: Butyrospermum Parkii (Shea) Oil, Persea Gratissima (Avocado) Oil, Brassica Campestris (Rapeseed Oil), Prunus Amygdalus Dulcis (Almond) Oil, Cocos Nucifera (Coconut) Oil, Avena Sativa (Oat) Oil

FATTY ALCOHOLS

Don't confuse fatty alcohols with alcohol. Fatty alcohols are a group of substances widely used in skincare. They're used to give a creamier and richer texture and have a moisurising effect. They also function as co-emulsifiers, meaning that they can help emulsifiers in skin cream to bind oils with other liquids. The positive side of fatty alcohols is that they can help to form so-called liquid crystals, which results in better moisturising properties.

INCI: Cetyl Alcohol, Cetearyl Alcohol, Behenyl Alcohol, Stearyl Alcohol

Johanna's Views on Natural Skincare

How many times have I been asked whether or not a cream is natural? I've lost count. When we use the word 'natural' today, what we often mean is that a substance has been extracted ready-made from a natural source. Examples are natural oils or butters from sunflowers or coconuts. But what's really most natural, I'd say, is to use the fats that already exist in our skin, the ones it's used to. Our skin doesn't produce sunflower oil, olive oil, or shea butter — but it does produce substances like squalene, ceramides, cholesterol, triglycerides, and many different kinds of fatty acids. Many skincare products are designed to imitate this unique combination of substances, partly by using natural oils containing the right combinations of lipids, and partly by extracting individual substances from oils and combining them in different ways. Quite simply, our skin's happiest with the substances it's used to, and they can either be extracted from natural sources or be synthesised.

Oils from the Plant Kingdom

OLIVE OIL consists mainly of oleic acid, with smaller amounts of other fatty acids such as omega 6 and palmitic acid. It's been found to contain over 200 different chemical compounds, including sterols, carotenoids, and phenol compounds. Phenols are olive oil's most abundant antioxidants, with more powerful antioxidant properties than vitamin E.

Olive oil has long been used in many cultures as an ingredient in skin and hair cosmetics. It's been shown to have anti-inflammatory properties and to play a role in healing wounds. However, its effect on the skin's barrier function isn't so good: it's been observed to increase TEWL in babies' skin, which can result in dehydration.

INCI: Olea Europaea (Olive) Oil

SUNFLOWER SEED OIL consists mainly of oleic acid and omega 6.
It has higher omega 6 levels than olive oil, making it better as a skincare ingredient. While sunflower seed oil contains fewer anti-inflammatory substances and antioxidants than olive oil, it has hydrating properties, probably thanks to the omega 6 it contains.

INCI: Helianthus Annuus (Sunflower) Seed Oil

RAPESEED OIL, which is very common in skincare products, consists of oleic acid, omega 6, and palmitic acid. While there are few studies of the impact of rapeseed oil on the skin, a comparison between a cream with 10 per cent glycerine and another with 20 per cent rapeseed oil showed that the former relieved skin irritation, while the latter made it worse. However, since it's rare for products on the market to contain as much as 20 per cent rapeseed oil, the value of this study is rather doubtful.

INCI: Canola Oil, Brassica Campestris (Rapeseed) Oil

PALM KERNEL OIL is used in 70 per cent of all cosmetic products on the market. This is because the yield per hectare of trees is very high, making this oil a profitable substance. Palm kernel oil contains palmitic acid, oleic acid, omega 6, and vitamin E. Relatively stable at high temperatures, it's odourless and has good emollient properties.
Unfortunately, the explosive increase in its use has resulted in massive deforestation of the rainforests. Given palm kernel oil's role in environmental destruction, many manufacturers are now trying to exclude any suppliers not certified as sustainable.

INCI: Elaeis Guineensis (Palm) Kernel Oil, Palmitate

COCONUT OIL is extracted from the kernel or flesh of ripe coconuts. It comprises many

GRAPESEED OIL is rich in phenol compounds, free fatty acids, and vitamins. It contains resveratrol, a phenol compound with strong antioxidant and anti-inflammatory properties. Grapeseed oil has been extensively researched. There's evidence that it improves healing and combats various pathogenic bacteria and fungi. As well as resveratrol, grapeseed oil has high levels of omega 6, vitamin E, and other phenol compounds.

INCI: Vitis Vinifera (Grapeseed) Oil

ARGAN OIL consists mainly of omega 9, palmitic acid, and omega 6 acids. It contains polyphenols, tocopherols, sterols, squalene, and triterpenic alcohols. Traditionally, argan oil has been used in cooking, treating skin infections, and in haircare products. Daily application of this oil has been shown to improve skin hydration by enhancing the skin's barrier function.

INCI: Argania Spinosa (Argan) Kernel Oil

fatty acids: lauric acid, myristic acid, palmitic acid, caprylic acid, capric acid, omega 9, omega 6, and stearic acid. A study of children with mild to moderately severe eczema (atopic dermatitis) showed that virgin coconut oil was effective in alleviating their condition. It's also been shown to have some effect in reducing inflammation. Its most important component, lauric acid, also helps combat pathogenic bacteria and fungi.

INCI: Cocos Nucifera (Coconut) Oil

SOYA BEAN (SOYBEAN) OIL contains genistein, a phyto-oestrogen whose chemical structure is similar to that of human oestrogen, making it a particularly valuable skincare ingredient. Genistein helps to replenish collagen, making the skin firmer. Soya bean oil has also been shown to protect against inflammation arising from sun exposure.

INCI: Glycine Soja (Soybean) Oil

SESAME OIL, which comes from sesame seeds, has traditionally been used in Taiwanese medicine to relieve pain from inflamed joints and scrapes or cuts. The seeds contain significant amounts of lignans such as sesamin, sesamolin, and sesaminol — substances with antioxidant effects.

INCI: Sesamum Indicum (Sesame) Seed Oil

WHAT'S IN A SKIN CREAM?

AVOCADO OIL contains omega 6, linoleic acid, and, above all, oleic acid. It also contains beta-sitosterol, beta-carotene, lecithin, minerals, and vitamins A, C, D, and E. This oil has some effect on collagen production and also has anti-inflammatory properties.

INCI: Persea Gratissima (Avocado) Oil

JOJOBA OIL, extracted from jojoba bush seeds, is widely used in skincare products. It's stable and doesn't turn rancid. Jojoba oil contains high levels of wax esters, making it effective as an emollient and humectant.

INCI: Simmondsia Chinensis (Jojoba) Seed Oil

BORAGE OIL, obtained from borage seeds, contains high levels of omega 6, which is important for the skin's barrier function. Studies in which children with atopic eczema wore T-shirts soaked in borage seed oil have shown that the omega 6 in the oil can relieve the condition. Borage seed oil has also been shown to boost the skin's barrier function and reduce TEWL.

INCI: Borago Officinalis (Borage) Seed Oil

APRICOT OIL, which comprises mainly omega 9, omega 6, and palmitic acid, comes from apricot kernels. Although it's very widely used in skincare, little scientific research has been conducted into this oil.

INCI: Prunus Armeniaca (Apricot) Kernel Oil

POMEGRANATE OIL, obtained from pomegranate seeds, consists mainly of omega 5 fatty acids and polyphenols. This oil is well known for its antioxidant and anti-inflammatory properties. Pomegranate extract has long been used to lighten the skin, as it inhibits melanocytes, the skin cells that produce pigment.

INCI: Punica Granatum (Pomegranate) Seed Oil

ALMOND OIL consists mainly of fatty acids: omega 9, omega 6, palmitic acid, and stearic acid. It has been shown to make stretch marks resulting from pregnancy less visible and to help prevent new ones from appearing.

INCI: Prunus Amygdalus Dulcis (Almond) Oil

CHAMOMILE OIL is remarkable for its high levels of bisabolol, used in many skincare products for its anti-inflammatory properties. It's been observed to help relieve itchy skin and eczema. However, chamomile oil on the skin has also been observed to cause contact allergy.

INCI: Anthemis Nobilis (Chamomile) Flower Oil

ROSEHIP OIL contains a large number of antioxidants, particularly tocopherols and carotenoids. The most common fatty acid is omega 6, followed by omega 3 and omega 9. Rosehip oil has strong antioxidant and anti-inflammatory properties. However, as with apricot oil, there are hardly any studies to show whether it really does have an effect on the skin.

INCI: Rosa Canina (Rose Hip) Seed Oil

OAT OIL consists mainly of omega 6 and omega 9 fatty acids. It's been used since time immemorial to treat a range of skin disorders such as rashes, reddening, burns, itching, and eczema. It's also been shown to have other positive properties, such as the capacity to boost ceramide levels in the skin.

INCI: Avena Sativa (Oat) Kernel Oil

WHAT'S IN A SKIN CREAM?

FATTY ESTERS

Fatty esters are a very common alternative to emollient oils in skin creams. This is because they're more stable than natural oils. Esters are derived from an alcohol or fatty alcohol plus a fatty acid. *Isopropyl palmitate*, for instance, is derived from isopropanol and palmitic acid.

Many fatty esters are used for their capacity to dissolve sun filters, which may otherwise have low solubility. In some cases, fatty esters also have a more pleasing cosmetic feel to them than pure natural oils.

INCI: Isostearyl Myristate, Alkyl Benzoate, Isostearyl Palmitate, Isostearyl Stearate, Isopropyl Palmitate, Butyl Stearate, Isopropyl Isostearate, Decyl Oleate, Isostearyl Neopentanoate

MINERAL OILS

Despite their name, there are no minerals in mineral oils. They have a bad reputation in the skincare world, and there's long been a debate as to whether they cause harm. Their critics say they form a film on the skin and stop it breathing. If mineral oils are used long term, it's claimed, they'll dry the skin so it becomes dependent on moisturising products. But there's no scientific proof for this — and even if it were true, it would apply equally to natural oils. My own view is that mineral oils are easy on the skin, provided that they meet high quality standards. Those that don't may contain traces of volatile substances that are harmful to the skin.

Unlike natural oils, mineral oils contain no unsaturated fatty acids or other nutrients.

Though this may sound like a drawback, the main point is that there's less risk of skin irritation. Unlike a natural oil, a mineral oil can never turn rancid. Most medical creams for dry skin or eczema contain paraffin, which is a mineral oil.

I don't want to suggest that mineral oils are always preferable in all situations. The picture's complicated. Environmentally speaking, it's better to avoid mineral oils since they originate from the petrochemical industry. However, it's not true to say they're worse for the skin than other oils and vegetable butters.

INCI: Mineral Oil Paraffinum Liquidum, Vaselin, Paraffin, Oleum Petrolen, Oleum Vaselini, Paraffinum, Vaselin Oil, Paraffin Oil, Liquid Paraffin

SILICONES

Silicones are the substances that give skincare products their silky texture. They're easily absorbed by the skin and help to hydrate it. Many products for oily skin or skin with acne are marketed as oil-free, and silicones are often used as the substitute.

These days it's hard to find any skincare products without silicones. Though classed as emollients, they're not the best of their kind — and it's not their emollient properties in skincare that make them so sought-after anyway. The reason for silicones' huge popularity with consumers is that they provide a quick fix. Applying silicone makes your skin feel silky straight away. Silicones are also handy for covering up uneven skin and even filling in small wrinkles.

The idea behind all this is to get consumers hooked on the pleasing sensation of applying silicones. I'm generally sceptical about adding a particular substance to a product just to give it a sensuous feel. This also cements preconceived ideas about how skincare products should feel on the skin — and that leads to counter-productive habits.

It's particularly bad when the additive also involves a health risk. In 2015, the Swedish pharmacy chain Apoteket banned all products containing cyclical silicones. This was because they've been shown to affect the liver and the respiratory organs if present in high concentrations.

Another problem with silicones is that they are non-biodegradable, which means they're harmful to the environment. Look out for names ending with –icone and –oxane in the INCI list, and beware!

INCI: Dimethicone, Cycloheptasiloxane, Cyclohexasiloxane, Cyclomethicone, Cyclopentasiloxane, Cyclotetrasiloxane, Cyclotrisiloxane

WHAT'S IN A SKIN CREAM?

Humectants

'Moisturise, moisturise, moisturise — you must have an effective moisturising cream!' says the shop assistant. If you've ever bought a skincare product, you've almost certainly been given this advice.

So what does moisturising — or hydrating — your skin actually mean? Does it help to take a shower? Sadly, the answer's no. For the skin to retain moisture, we have to add an ingredient that binds water in the skin. That's where humectants come in.

GLYCERINE

Glycerine, also known as glycerol, is found naturally in the skin, where it binds water in the topmost layer and hydrates the skin. The most common humectant in skin creams, it's derived from vegetable fats. In addition to its humectant properties, glycerine has been shown to reduce skin irritation.

Medical products to treat very dry skin use high concentrations of glycerine — as much as 10 to 20 per cent.

However, too much glycerine can leave a sticky after-feel when applying the product. The Swedish word *lagom*, which means 'moderate', is a good word when it comes to glycerine. Using products with less than 5 per cent glycerine will help you avoid that uncomfortable after-feel.

INCI: Glycerol, Glycerine

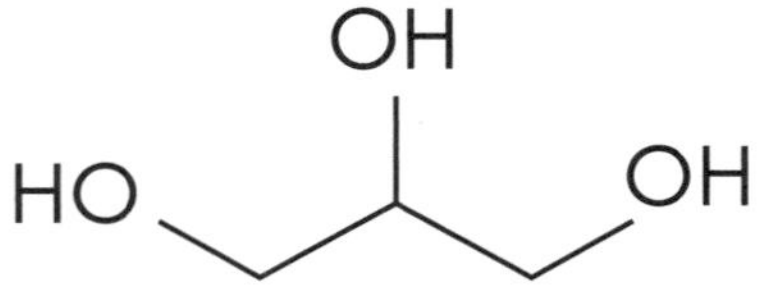

UREA

Another common humectant is urea, which is found naturally in urine. Urea (sometimes known as carbamide) is used in medical skincare products to treat dry skin. Added urea often accounts for between 2 per cent and 10 per cent of the volume. If the concentration is any higher, the cream may sting slightly on application. Urine has actually been used for a long time to treat dry skin and eczema, as Ancient Greek and Roman texts show. Synthetic urea production began in the nineteenth century.

INCI: Urea, Carbamide

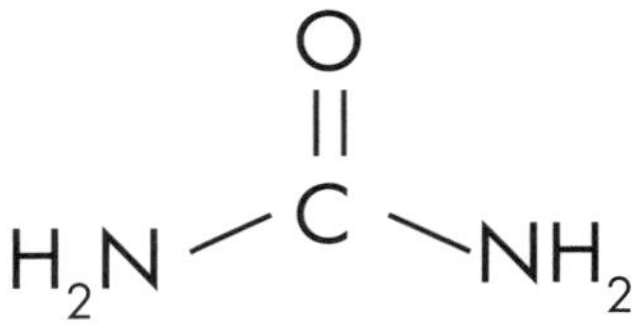

UREA

GLYCERINE

HYALURONIC ACID

These days it's hard to find a skincare product that doesn't contain any hyaluronic acid. This substance, which occurs naturally in our skin, has an amazing capacity to bind water. In fact, it can bind up to a thousand times its own weight of water. If injected, hyaluronic acid binds the water naturally present in the body, so it can be used to plump up lips, fill in wrinkles, and enlarge breasts.

However, applying hyaluronic acid externally has very little effect. Although a piece of beefsteak will provide you with nutrients and amino acids once it's been chewed and swallowed and made its way through your digestive system, patting your skin with a steak will have precious little effect on your gut. Similarly, the hyaluronic acid molecule is too large to be able to penetrate even the topmost layer of your skin. More on molecular size on p. 68.

In recent years, manufacturers have started producing hyaluronic acid with smaller molecules (LMWH). Unfortunately, even the LMWH molecule is too big to be able to get through the stratum corneum. Although skincare manufacturers sometimes label products to show they contain LMWH, it's generally hard to identify it from the ingredients list. The usual term is *sodium hyaluronate* — which tells us nothing about molecule size.

Summing up the research situation, there's little scientific evidence that using either hyaluronic acid or LMWH in skincare does anything much to hydrate the skin. The clinical studies available used other humectants, such as glycerine, in a ratio of 1 per cent hyaluronic acid to 5 per cent glycerine. That means they tell us little about hyaluronic acid as such.

INCI: Sodium Hyaluronate, Hyaluronic Acid

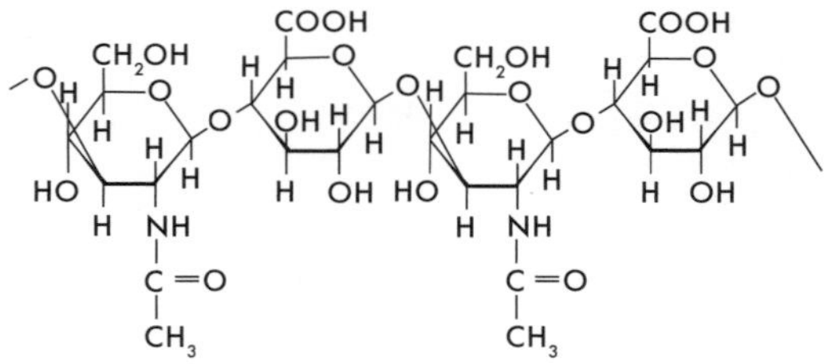

HYALURONIC ACID

Hyaluronic acid (structure depicted above) is a large molecule, so there's no way it can get through the stratum corneum.

BUILDING BLOCKS OF HYALURONIC ACID

Instead of hyaluronic acid or LMWH, it's better to use one of the chemical building blocks that make up this acid. The skin contains an enzyme that produces the variant that occurs naturally there. That enzyme's precursor, *N-acetylglucosamine*, is better able to penetrate the surface of the skin. Once in the skin, it helps to produce hyaluronic acid, which can then go on to bind water and hydrate the skin.

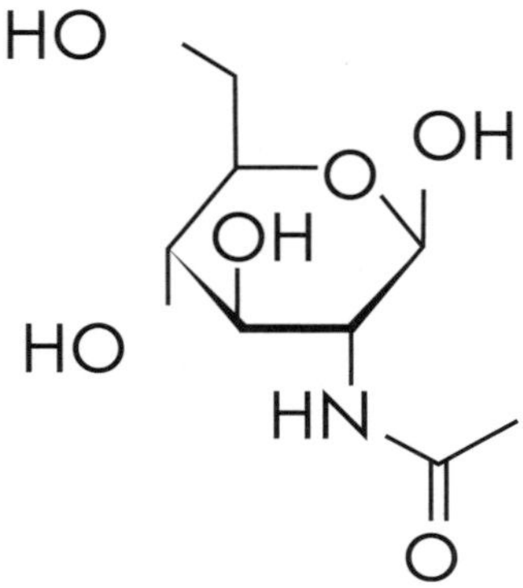

N-ACETYLGLUCOSAMINE

Skincare Regulation

Skincare manufacturers are responsible for making sure their products are safe to use. This is done in two ways:

1. In skincare, microbiological stability means making sure the product can't provide a breeding ground for bacteria and fungi. This means a challenge test has to be carried out before a product can be put on the market. Such a test involves adding at least four different kinds of pathogenic bacteria and fungi to a face cream — a million of each type to each gram of cream. For the cream to pass the test, the bacteria and fungi must dwindle within a few days. Meeting this challenge calls for high levels of preservative or other antibacterial substances.

These tests are conducted to make sure that skincare products can be kept for at least thirty months and stored at room temperature (or even higher temperatures). Quite simply, it makes economic sense to produce large volumes of the product and keep it in storage, instead of producing smaller quantities at shorter intervals.

Just imagine placing the same level of demands on food. If similar requirements applied, customers would be able to keep crème fraîche or milk at room temperature, or in a warm bathroom, for two and a half years without getting a bacterial or yeast overgrowth. That gives you some idea of how much preservative is needed to keep the skin cream intact.

2. The second safety requirement is that the product must not harm the user's health. This means that the manufacturer must not use any substances proven to cause allergic reactions. But there's some inconsistency here. Most preservatives can cause allergic reactions, and so can other substances commonly used in skincare products.

ANIMAL TESTING

The testing of cosmetics (and their individual ingredients) on animals is banned in Europe. In China, however, animal testing is mandatory. Skincare products that haven't been tested on animals aren't allowed into the country. Many skincare companies that export worldwide therefore choose to compromise — even if their official policy bans animal testing. Given the size of the Chinese market, staying out of it is a hard economic choice. But the more companies abide fully by their own policies, the better it will be for animal rights.

AMINO ACIDS

Amino acids, found naturally in our skin, act as humectants.

> **INCI:** Serine, Pyroglutamic Acid, Pyrrolidone Carboxylic Acid (PCA) Alanine, Glycine, Lycine, Threonine, Arginine, Proline

SALTS

Salts are also natural humectants.

> **INCI:** Sodium Chloride, Magnesium Chloride, Sodium PCA

GLYCOLS

Glycols are solvents used frequently and in large amounts in skincare. This places them high up on the INCI list. This doesn't necessarily mean they're the best humectants available; they are also preservatives, just like alcohol.

In addition to being humectants and preservatives, glycols act as penetration enhancers. This means they enable active ingredients to enter the skin more easily. This property also means that glycols don't block the skin's pores, which makes them good for people with oily skin.

There is, however, a downside to widespread glycol use; high levels of some glycols on the skin risk causing a negative skin reaction of some kind. In fact, propylene glycol had the dubious honour of being named 'allergen of the year' by the American Contact Dermatitis Society in 2018. They can cause skin irritation that can manifest as bumps or spots. Up to 3.5 per cent of everyone alive is allergic to propylenglycol. In 2018, the famous Mayo Clinic in Rochester, USA, conducted a major study on the subject. The clinic concluded that there was a link between glycol and skin reactions or allergies. Accordingly, it advises consumers to exercise caution. Glycols are easily recognisable in the list of product ingredients now, because they always end in –glycol or –diol.

> **INCI:** Butylenglycol, Pentylenglycol, Propylenglycol, Propanediol (1,3 Propylenglycol), Butanediol (Butylenglycol), Dipropylenglycol, Ethylenglycol, 1,2 Pentadiol

WHAT'S IN A SKIN CREAM?

Active Ingredients

Active ingredients rejuvenate, hydrate, and smooth out wrinkles and uneven skin tone. But they only work if present in high-enough concentrations, and if they're properly protected throughout the production–transport–storage–sale process, and all the way home to you, the consumer. You generally need to use active ingredients for at least three months for them to have any effect.

Many active ingredients are sensitive to moisture, air, light, and water. These factors cause them to break down and become ineffective in daily use.

One of the main challenges to functional skincare is the way we produce and store our products. Packaging them in glass jars with lids we can open daily — hello, light! — and storing products for long periods under bright lights in shops — hello, even more light and heat! —are two of the highly effective techniques we've devised to destroy face creams' active properties.

The skincare industry could quite easily take steps to ensure that active ingredients stay active until you get them home. The way to do this would be to reduce the shelf life of skincare products, develop products requiring cold storage, use nitrogen in the production process (to exclude oxygen from the packaging), or make water-free products to which users could simply add water before use. But at the moment, manufacturers and consumers are stuck in a compromise that's far from ideal, for either the skin or product development.

MOLECULAR SIZE MATTERS

The skincare industry likes to refer to in vitro studies of skin cells (fibroblasts, keratinocytes, or melanocytes) as evidence of how effective particular skincare ingredients are. But these tests tell us little about what effect an ingredient has if you apply it directly to the skin. What matters is whether the molecule is small enough to penetrate the top layer of skin, so it can reach the cells it's intended for.

Molecular size is measured in units called daltons. The usual guideline is that a molecule has to be about 500 daltons or smaller for the substance to penetrate the top layer of skin in the first place. It should also preferably be lipophilic (fat-soluble) to get through the stratum corneum.

Retinol, for example, has relatively good penetrative properties, as it's fat-soluble and has a molecular size of 286 daltons. The more stable variants of retinol, such as retinol palmitate, have a molecular size of 524 daltons. It's harder for these larger molecules to pass through the top layer of skin.

Vitamin C is also a small molecule, at 176 daltons. But it's water-soluble, so a higher concentration is needed to have an impact.

Compare these with hyaluronic acid, whose molecule measures about 1,000,000 daltons and has absolutely no chance of penetrating the skin. However, there's a variant of hyaluronic acid with a low molecular weight of about 50,000 daltons. There's also a variant with an ultra-low molecular weight of just 5000 daltons. Yet even this is still ten times too large to penetrate the stratum corneum. Collagen molecules measure 300,000 daltons, which means they can't have any effect at all on the skin if applied externally.

In this context, it's interesting to note that phenoxyethanol, a very common

ABOUT SKINCARE

Ask Johanna:

Can retinol have side-effects?

Some individuals are oversensitive to retinol, and in them, it can cause redness and irritation. In the course of clinical studies, I've noticed myself that participants are sometimes more sensitive to retinol in winter. Other studies have confirmed this link.

preservative, has a molecular size of 138 daltons, which means it can easily penetrate the skin. The same is true of parabens, whose molecular size varies between 150 and 200 daltons. Let's bear in mind that both of these additives serve to protect the product — not our skin.

RETINOL

Retinol is one of the best-studied substances in dermatology and the cosmetics industry. It's also among the few substances proven by clinical studies to have an impact on the skin. Retinol is also the active ingredient that's been most heavily marketed over the last decade. In chemical terms, retinol is an alcohol, and it's synonymous with vitamin A.

The importance of retinol to the skin emerged during World War I, when it was discovered that a deficiency of the substance thickened and hardened skin and made it very dry. Modern skin treatments with retinol began in 1968, when it was first synthesised. In the last two decades, it's also been used to treat various skin disorders, including acne.

It was through treating skin diseases that retinol's 'rejuvenating properties' were discovered. Patients who used it developed fewer and finer facial lines, and their skin became more supple. Later it was discovered

that retinol boosts collagen and elastin production, while also helping to make the surface of the skin more even, resulting in smoother skin with fewer wrinkles.

It can be particularly beneficial to add retinol to night cream. This is because exposure to UV rays increases the amount of retinol naturally present in the skin. If the concentration is too high, retinol can irritate the skin, particularly during the coldest six months of the year, when our skin's often a little more sensitive.

But do bear in mind that retinol is also sensitive to air and temperature, meaning that it can easily be destroyed. Never buy a cream marketed for its retinol content if it's sold in a jar. Taking the lid off regularly will make any retinol in the product significantly less effective.

RETINOL

INCI

The kind of retinol that counts appears on the INCI list simply as 'Retinol'. There are many more-stable variants of the substance, but most are less effective than the retinol that occurs naturally in our bodies. Variants of retinol always begin with 'retinyl'. Retinyl acetate and retinyl palmitate are examples.

NIACINAMIDE

These days, thousands of products list niacinamide among their ingredients. Niacinamide is the active ingredient on which most clinical studies have been carried out, and it has many effects: it enhances the skin's barrier function and reduces TEWL, mitigates irregular pigmentation, and curbs the production of sebum by the sebaceous glands. These qualities make it beneficial to both oily and dry skin. To be effective, however, it has to be over 2 per cent of the skincare product concerned.

VITAMIN E

Vitamin E, also known as tocopherol, is an antioxidant widely used in skincare that's been shown to have many positive effects on, for example, extremely dry skin, eczema, superficial burns, lesions, and other skin disorders triggered by inflammation. Scientists have also shown that this vitamin has some impact on damage caused by exposure to the sun. Being fat-soluble, vitamin E has to be dissolved in an oil phase.

INCI: Tocopherol and more stable variants such as Tocopheryl Acetate

VITAMIN C

Vitamin C, an antioxidant, is marketed for its capacity to reduce wrinkles, remove patches of pigment, boost elasticity, and even out the user's skin tone. It's water-soluble and has been shown to have positive effects when applied directly to cells in the lab.

Being hydrophilic (water-soluble) means it's hard for the molecule to penetrate the skin. A fairly high concentration is needed for it to have any effect. At suitably high concentrations, vitamin C has been shown to have a relatively beneficial effect on ageing skin, resulting in fewer wrinkles. But this doesn't happen overnight; the treatment needs to be continued for three months or so.

It would be a good idea to put vitamin C in water-free products. This would enable it to penetrate the skin and would also extend its shelf life.

INCI: Vitamin C, Ascorbic Acid, Ascorbyl-6-Palmitate, Magnesium Ascorbyl Phosphate, Ascorbyl Glucoside and other ingredients including the word 'Ascorbyl'.

VITAMIN C

PLANT STEM CELLS

Over the last few years, plant stem cells have become an increasingly common ingredient in skincare. The effects vary depending on the plant from which the stem cells are derived. There are stem cells from roses, lilac, algae, edelweiss, and numerous other plants. Some boost collagen formation, while others have antioxidant properties or help to control sebum.

Stem cells are obtained by scraping them off the plant and cultivating them in the lab. Once there's a sufficiently large amount, all you have to do is extract the active substances and use them in skincare.

Skincare products based on plant stem cells have various advantages. One is the avoidance of land use to grow plants for the skincare industry, freeing it up to grow food instead. A further advantage is that stem cells have the unique capacity to grow indefinitely.

Stem cells from human sources are never used except for medical purposes. They are extracted from bone marrow, mainly for use against different kinds of cancer.

INCI: Plant + callus or cell culture extract example: Aloe Vera Callus Extract

WHAT'S IN A SKIN CREAM?

Acids

Have you noticed the terms AHA, BHA, or PHA on packaging? All three are acids. There are hundreds of different types of acid, but when we talk about acids in the context of skincare, we nearly always mean one of these three. Their main purpose is exfoliation, which gives the skin a more even texture and tone.

AHA — THE MOTHER OF ALL ACIDS

The most common AHA acids are glycolic and lactic acid, which work on the keratinocytes, thinning the top layer of skin. Acne sufferers may experience a reduction in pore size and the number of infected blackheads. This is due to the impact the acid has on the keratinocytes and has less to do with any effect on the secretion of sebum and oils. The acid is said to have a *keratolytic* effect.

Acids can also lighten patches of pigment. They do this by speeding up the process of evening out the keratinocytes, enabling the melanin formed in the basal layer of the epidermis with the help of the melanocytes to move towards the surface, where it is removed in dead skin cells.

Both glycolic and lactic acid can cause irritation, owing to their low pH values. They can also affect the TRPV1 receptor, resulting in irritation.

INCI: Lactic Acid, Glycolic Acid, Malic Acid

GLYCOLIC ACID

BHA — NOTHING NEW UNDER THE SUN

Salicylic acid is the most widely used BHA acid. The first studies on its effects in skincare were carried out in the 1950s as part of research into psoriasis. It wasn't until much later that its cosmetic effects on wrinkles and hyper-pigmentation were observed. Salicylic acid, which is derived from willow trees, has been used for 2000 years to treat various kinds of skin disorders.

Today it can be found in everything from face creams and foot creams to cleansers. BHA acids have the same impact on pore size as AHA acids. The difference is that salicylic acid is slightly more lipophilic (fat-soluble), so it enters the skin more easily. As a result, it penetrates more deeply into the sebaceous glands, enabling it to prevent the formation of blackheads and whiteheads.

The necessary concentration interval for salicylic acid to be effective is 1.5 per cent to 3 per cent, but the proportion in chemical peels can be as high as 30 per cent. Salicylic acid can irritate the skin. One tried-and-tested way to lessen the irritant effect of AHA and BHA is to raise the pH level of the skincare product concerned.

INCI: Salicylic Acid, Willow Bark Extract

SALICYLIC ACID

ABOUT SKINCARE

PHA — THE LATEST ADDITIVE

The difference between PHA on the one hand and AHA and BHA on the other is their respective molecular structures. Gluconolactone and lactobionic acid are two examples of PHA acids. Both are glycated acids with a humectant effect on the skin. PHA acids are among those least likely to cause irritation. People with sensitive skin can generally tolerate them.

INCI: Gluconolactone, lactobionic acid

GLUCONOLACTONE

Peptides

Peptides, often mentioned in sales pitches for skincare products, consist of a chain of amino acids. They can achieve remarkable results; for instance, they can both boost collagen production and soothe the skin. The tests behind the marketing have shown how peptides affect cells cultivated in vitro, but not how they affect actual skin. However, peptides with a relatively low molecular weight (those comprising a small number of amino acids) can penetrate the skin and reach the relevant cells.

> ### BE CAREFUL WITH ACIDS DURING MONTHS WITH HIGH UV EXPOSURE
>
> Since AHA and BHA have an exfoliating effect (they remove dead skin cells), which makes the skin thinner, this treatment also increases sensitivity to sunlight. Paradoxically, many people experience an increase in pigment patches when undergoing acid treatment. This is because it's easier for the sun to reach the pigment cells, triggering over-pigmentation.
>
> PHA, on the other hand, doesn't increase sensitivity to this extent.

Collagen

In the last few years, collagen has been hyped as one of the real superstars of skincare. But I've got one thing to say about collagen in skincare products — in a skin cream, pure collagen will have no effect whatsoever. Collagen molecules comprise over 1000 amino acids, making it impossible for them to enter the skin at any point. While hydrolysed collagen has a lower molecular weight, molecules still measure between 1000 and 5000 daltons, which means they have very little chance of reaching the fibroblasts in the dermis.

If products are labelled as 'procollagen' or 'collagen-boosting', on the other hand, it's a different story. Here we're talking not about collagen itself, but about substances that can stimulate the skin to produce collagen of its own — and that actually *does* work.

COLLAGEN

Antioxidants

Antioxidants of various kinds are used in nearly all skincare these days. We've already mentioned some of them, including vitamins C and E. Many antioxidants are good for the skin. They slow down the breakdown of collagen and elastin, making the skin more supple and reducing wrinkle formation. For this to happen, it's again important to have the right concentration of antioxidants, and for them to be fully absorbed by the skin.

We know that sun ages skin more than anything else. UV light increases both inflammation and levels of the MMP enzyme that breaks down collagen and elastin. The pigment-secreting cells, melanocytes, are also triggered so that they produce different amounts of pigment in different areas of skin, resulting in pigment patches. Antioxidants, if they've been correctly stored, have the capacity to break down free radicals and reduce pigment formation.

Antioxidants are unstable substances that break down easily at high temperatures. Remember the spotlights in the cosmetics boutique — how warm does it get there? That's why nearly all antioxidants used for research purposes are kept in a fridge or freezer. They also break down on contact with oxygen.

The antioxidants on which most studies have been conducted are vitamin C, ferulic acid, lipoic acid, vitamin E, and — last but not least — epigallocatechin-3-gallate, which gives green tea its antioxidant properties.

INCI: Ferulic Acid, Ascorbic Acid, Lipoic Acid, Tocopherol, Epigallocatechin-3-Gallate

South Korean Ingredients

The market for South Korean skincare products is growing apace. The South Korean skincare routine often involves over ten products: everything from creams, serum, serum sprays, and micellar water to sheet masks. This may give the impression that South Koreans have a superior, more sophisticated skincare regime. However, using more products also means putting more of the less beneficial substances on your skin: more preservatives, more perfume, more glycols, and so on.

Cosmetic ingredients are certainly rigorously tested in South Korea (mainly through animal tests, though legislation in this area is changing at the time of writing), and a great deal of effort has been invested in developing new product types. Much emphasis is placed on creams with a special, sensuous feel to them. Many companies have special labs whose sole function is to assess a product's sensory qualities, i.e. how it feels on your face.

South Korean skincare products contain more or less the same ingredients as contemporary skincare products from the rest of the world. In 2017, US dermatologists Quay, Chang, and Graber from the Boston Medical Center's Department of Dermatology carried out a wide-ranging study of Korean skincare and popular products such as liquorice, niacinamide, snail slime, ginkgo biloba, ginseng, green tea, pomegranates, and soya. Of the various ingredients assessed, only niacinamide, green tea, liquorice, and soy proved to show some genuine anti-ageing effect. The study concludes that the evidence for the effectiveness of most of the other ingredients is limited.

TIP

When you're selecting a skin cream, I think the main thing you need to ask yourself is this: what do skin cells need in their natural state — that is, in the skin? The answer's simple — specific amino acids, vitamins, and minerals.

Amino Acids

Amino acids are a subject close to my heart. I've been involved in research into the role they play in the skin for the last decade. Skin cells need amino acids to build proteins such as collagen and elastin.

In one project, we also observed that a particular amino acid normally found in the brain, N-acetyl aspartic acid, had a powerful effect on the firmness of the skin. After twelve days of treatment, we observed a significant increase in the levels of both collagen and elastin in the skin. The results were so good that we took out a worldwide patent on the substance.

INCI: Glycine, Alanine, Arginine, Asparagine, Aspartic Acid, Acetyl Aspartic Acid, Cysteine, Glutamic Acid, Glutamine, Histidine, Isoleucine, Leucine, Lysine, Methionine, Phenylalanine, Proline, Serine, Threonine, Tryptophan, Tyrosine, Valine

Minerals

Minerals are good for the skin because they activate many important enzymes. Selenium, for example, activates several enzymes that make up important antioxidant systems and play a significant role in the growth of skin cells.

Zinc, manganese, and copper are other examples of minerals, and, by activating the antioxidant enzyme superoxide dismutase (SOD), they can break down more harmful oxygen radicals into hydrogen peroxide. Hydrogen peroxide is a little better than free oxygen radicals, but still dangerous to the skin. An enzyme called catalase comes to the rescue here, breaking down hydrogen peroxide into water and pure oxygen — an ideal final product. Catalase is activated by manganese, another mineral.

Part of my doctoral research involved putting together a clinical sample of Dead Sea water for the treatment of vitiligo. Dead Sea water contains the following minerals: magnesium, calcium, sodium, potassium, selenium, and manganese. While ordinary seawater contains the same minerals, they are present in far smaller quantities. For the body to be able to absorb these minerals, they must be available in the form of salts.

Skincare products contain minerals in countless different forms, including the examples below.

INCI: Sodium PCA, Magnesium PCA, Zinc PCA, Manganese PCA, Copper PCA, Manganese Gluconate, Calcium Gluconate

WHAT'S IN A SKIN CREAM?

Story Ingredients

'Story ingredients' is a term widely used in the skincare industry. These magic ingredients are added to products to create a good story; in other words, they're just part of the sales pitch. The idea is to add an ingredient that sounds attractive — preferably something natural, or just an ingredient that sounds expensive. Special claims are made for this ingredient, which could be anything from diamond powder to caviar extract.

A product marketed as containing caviar supposedly has the following effects:

Lifts and firms skin.
Helps restore the appearance
of elasticity and tone.
Smooths and refines skin texture.

The downside for consumers is that the special ingredients are present only in extremely low concentrations — so low, indeed, that tests show they have no effect at all — so we risk paying through the nose for ingredients that add nothing useful to the product.

Take a closer look at the INCI list and you might see that this product does contain some caviar extract, but it also contains more stable variants of active substances such as retinol and vitamin C, which have precisely the effects attributed to caviar. However, it's now illegal to market products on the basis of specific ingredients and their putative effects. Instead, the way a product is marketed must reflect its overall qualities.

Preservatives

Have you ever seen the date of manufacture on a skincare product? No?
That's just as I thought. It's hardly ever indicated.

Your skin cream contains preservatives of various kinds to enable it to keep for at least 30 months, and to kill bacteria and fungi. Creams with a long shelf life can be transported over long distances and stored for long periods.

Products with a high water content contain particularly high levels of preservatives, as water is a breeding ground for bacteria and fungi. As well as killing any bacteria that might start to grow in your skin cream, preservatives can also affect the skin's good bacteria once the cream is applied to your skin. This could potentially lead to an imbalance in the skin microbiome, which might result in one of several common skin disorders. We'll take a closer look at this in the chapter on the microbiome.

Manufacturers have a duty to guarantee the safety and quality of skincare products, but there's something wrong when they focus mainly on a long shelf life, instead of on finding new approaches to distribution and logistics that might result in a new, better kind of skincare.

WHEN PARABENS WERE BLACKLISTED …

Parabens are among the skincare sector's most frequently blacklisted substances. If a product is paraben-free, you can guarantee its packaging will flag that prominently.

In 2004, scientist Philippa Darbre published an article in the UK on a result observed by her research team: parabens had been seen to accelerate the growth of breast tumour cells. The fact that the structure of some parabens resembled that of oestrogen suggested that they might be linked with the onset of breast cancer in

women. It's still not clear how much paraben can be absorbed into the bloodstream through the skin, and the study hasn't been verified, but as the rumour reached consumers worldwide, many manufacturers chose to remove parabens from their products. The most widely used parabens, methylparaben and ethylparaben, are still considered safe in skincare. In fact, parabens have a very low frequency of reported contact allergy. Since 2017, it's been illegal in the EU to label a product as 'paraben-free', a description that was very popular for many years.

> **INCI:** Methylparaben, Ethylparaben, Methyl p-hydroxybenzoate, Methyl Parahydroxybenzoate, Ethyl Parahydroxybenzoate, Ethyl Parahydroxybenzoate, Ethyl Parahydroxybenzoate, Ethyl p-hydroxybenzoate, 4-Hydroxybenzoic Acid Ethyl Ester

… A NEW PLAYER APPEARED ON THE SCENE

Parabens made way for a different preservative, phenoxyethanol. Today this is by far the most common preservative used in skincare products. A number of studies have shown a link with contact eczema and allergies. Moreover, manufacturers that have replaced parabens by phenoxyethanol have received more complaints about allergic reactions. There's even talk of an epidemic of allergies. Scientists have actually shown that phenoxyethanol triggers the same receptor as chili peppers, TRPV1.

The French authorities have recommended that phenoxyethanol should not be added to products intended for use with nappies (diapers) for the under-threes, yet the substance is still in use in Sweden.

> **INCI:** Phenoxyethanol

WHAT'S IN A SKIN CREAM?

... AND THERE'S A THIRD PRESERVATIVE THAT'S UNPRONOUNCEABLE

The preservative methylisothiazolinone is used as an alternative to parabens and phenoxyethanol. Methylisothiazolinone releases formaldehyde, recognised as an allergenic substance, onto the skin. It's no exaggeration to speak of an epidemic of allergic reactions: this additive produces reactions even in extremely low concentrations, such as 0.0015 per cent.

Recently the EU decided that methylisothiazolinone must no longer be used in leave-on products (products that aren't rinsed off immediately after use). Fortunately, that means the substance is now being phased out of skincare products.

Generally, EU cosmetic legislation serves as an inspiration for the rest of the world, and when an ingredient gets banned, other countries tend to follow quickly, including Latin America, the ASEAN region, and others. In Australia, it is banned, too. However, it's not banned for use in the US, and the FDA do not have such a strict approach to ingredient restrictions — they don't have annexes of banned and restricted ingredients like the EU does, and follow their own, more lenient approach to cosmetic regulation.

Other INCI substances that release formaldehyde: 2-Bromo-2-nitropropane-1, 3-Diol, Diazolidinyl Urea, DMDM Hydantoin, Imidazolidinyl Urea Quaternium-15

To date, there are sixty allowed preservative systems for cosmetic products. Some of the most common ones are listed here.

INCI: Phenoxyethanol, Benzoic acid, Sodium benzoate, Salicylic acid (Salicylate), Sorbic acid, Potassium sorbate, Ethylhexylglycerin

PRESERVATIVE-FREE — WHAT DOES THAT MEAN?

To be able to label their skin creams as 'preservative-free', many manufacturers use substances that aren't classed as preservatives — alcohols, glycols, or other solvents — but which have the same function.

INCI: Alcohol, Propylene Glycol, Pentylene Glycol (and other substances ending in —glycol)

ESSENTIAL OILS

Organic products often use essential oils instead of conventional preservatives for their antibacterial qualities. Essential oils typically have a pungent scent and they frequently contain limonene and linalool, which can have a highly irritant effect on the skin.

It's also been shown recently that they can be phototoxic, sometimes producing a strong inflammatory reaction in the skin on contact with sunlight.

INCI: Clary Sage Essential Oil, Eucalyptus Essential Oil, Geranium Essential Oil, Ginger Essential Oil, Jasmine Essential Oil, Lemongrass Essential Oil, Neroli Essential Oil, Oregano Essential Oil, Patchouli Essential Oil, Rosemary Essential Oil, Sage Essential Oil, Sandalwood Essential Oil

Innovation and the Future

There are ways to minimise preservative use. One is to make the manufacturing process cleaner. This would make it unnecessary to use preservatives to protect the product from bacteria and fungi during manufacturing. Another is to use fresher (shorter-lived) products, while also introducing shorter expiry dates. Third, more hygienic pack-aging options. Please say goodbye to jars; airless pumps are the best option.

There's also interesting research and development under way on pasteurisation, with a view to making skincare products cleaner, safer, and longer-lasting. I hope we'll soon see more innovation in manufacturing, distribution, and storage.

Ask Johanna:

What about organic products?

Organic products can contain preservatives too. Preservatives approved by the organic certification organisation Ecocert are:

INCI: Potassium Sorbate, Sodium Benzoate, Benzyl Alcohol, Benzoic Acid, Sorbic Acid, Dehydroacetic Acid, Sodium Dehydroacetate

Emulsifiers

Emulsifiers are present in all emulsions made up of fats and water that have to be held together over time. They are a necessary ingredient in conventional skincare products.

By far the most common emulsifier today is polyethylene glycol (PEG) in its various forms, combined with fatty alcohols. PEGs are not used in natural or organic products. Instead, these use alternatives such as sorbitan olivate, glyceryl stearate citrate, and cetearyl olivate. There are also other types of emulsifiers that rely on particles to mix oil and water, forming what's known as a Pickering emulsion.

INCI: PEG-s (PEG-100 Stearate, PEG-40 Stearate,PEG-100 Lanolin, PEG-10 Glyceryl Diisostearate, PEG-10 Castor Oil), Sodium Laureth Sulphate), Polyglyceryl Oleate, Sorbitan Olivate, Glyceryl Stearate Citrate, Cetearyl Olivate, Ceteareth-20, Polysorbate 80

HOW ABOUT WATER-FREE PRODUCTS INSTEAD?

Water-free creams and salves are less likely to provide a breeding ground for bacteria and fungi. This is because bacteria and fungi grow in water, but not in oil. What's less good about an oil-based cream is the lack of a water phase to dissolve water-soluble ingredients — and most skincare ingredients are water-soluble. Substances that don't combine well with water-free creams include niacinamide, vitamin C, amino acids, minerals, and, last but not least, all humectants (such as glycerine and urea).

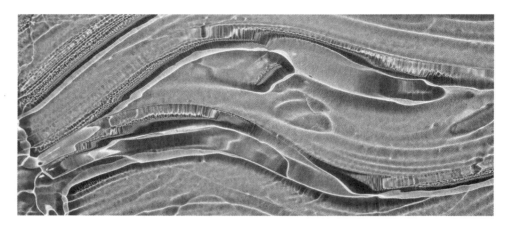

Thickeners and Gelling Agents

Your skin cream contains various thickeners and gelling agents to make it easier to apply but not too greasy (like a salve). Thickeners include xanthan gum, carbomer, and other polymers. They're mostly gentle on the skin. If the thickening agent is a dominant ingredient in the cream, it may make your skin feel quite sticky. There's nothing wrong with that as such, but it isn't a particularly agreeable sensation.

Acrylamide polymers are frequently used as thickeners. The group comprises acrylates, acrylamides, methacrylates, and other acrylic polymers. In addition to acting as thickening agents, they form a coating on the skin, giving your skin cream a pleasant texture. IHowever, some acrylamide-based polymers are associated with microplastics and are very hard to break down.

To make a gel, you need only water and a gelling agent. These form a hydrogel. If you add a little oil, you get an oil gel. These formulations are mostly used on fairly greasy skin. They may be entirely emulsifier-free.

Thankfully, development in this area is ongoing, and there are now many alternatives for biodegradable thickeners. Natural gums such as sclerotium gum, Caesalpinia Spinosa Gum, xanthan gum,

and pullulan, or thickeners based on cellulose as a thickener and film-forming agent. Microcellulose from wood and other sources is already used in skincare, but what's new is nanocellulose, a fascinating form of the substance. Cellulose is biodegradable.

INCI: Acrylamide-based polymers: Polyacrylic Acid, acrylates / C10–30 Alkyl Acrylate Crosspolymer, Hydroxyethyl Acrylate / Sodium Acryloyldimethyl Taurate Copolymer, Polymethyl Methacrylate, Polyacrylamide

Natural gums: Xanthan Gum, Sclerotium Gum, Caesalpinia Spinosa Gum, Pullulan

Cellulose-based polymers: Microcrystalline Cellulose, Cellulose Gum

Masking Agents

Fragrances can take us to unexpected places and conjure up remembered feelings. When I go into a spa or breathe in the aroma of freshly brewed coffee, a sense of calm comes over me. When fragrances are used in skincare products, however, they may cause allergic reactions.

Masking agents are added to a cream to mask an unwanted odour or colour.

PERFUMES

Perfume is added to your skin cream to make you feel good, but also to cover up any unwanted odour. Unperfumed skincare products can fall anywhere on the scale between odourless and unpleasant smelling.

Most consumers seem to prefer having perfume in their skincare products. I once met a woman who was terrified that the night cream she used might disappear from the product range. Her husband adored the scent of her face, and she was convinced that he loved her only for her night cream.

But what's wrong with perfume? Well, many substances used in perfumes are allergenic. The EU has compiled a list of twenty-six allergens found in perfumes. Some of the most common are citral, farnesol, limonene, and linalool — precisely the ones widely used in skincare.

If you must use perfume, it's better to spray your favourite fragrance in your hair or to apply a drop or two to your neck — but avoid using perfumed products on your face, especially if you have sensitive skin.

Emollient oils have no smell of their own, at least not one that's fresh or attractive. If there is a fragrance, it comes from essential oils, which can also cause an allergic reaction.

INCI: Citral, Farnesol, Limonene, Linalool, Clary Sage Essential Oil, Eucalyptus Essential Oil, Geranium Essential Oil, Ginger Essential Oil, Jasmine Essential Oil, Lemongrass Essential Oil, Sandalwood Essential Oil, etc

COLORANTS

Ever bought a skincare cream with pink flowers on the pack, in which the product is also pink? Many people find this positive, assuming there is a high volume of that ingredient. Sorry to disappoint you — most often, colorants are used.

Colorants are also added to mask the discolouration that occurs as a cream oxidises. They may also be used to give the product a more yellowish tone, which users perceive as being natural. One of the colorants most often used in skincare products is iron oxide.

Colorants are generally listed by their Colour Index (CI) number.

INCI: CI 77492, CI 45410, CI 19140, CI 45380

WHAT'S IN A SKIN CREAM?

Skincare That Works

I remember the face cream my mother used when I was a child. It came out of a white tube with ornate gold lettering. She applied it every morning and evening, and in summer she'd add sunscreen. At the time (the early 1980s), skincare products were still a more or less exclusively female prerogative. Most women had just one or two of them.

Since then, the range available on the market has exploded.

Today we face a plethora of night creams, day creams, eye contour creams, UV screen products, serums, tone-up creams, emulsions, lotions, essences, boosters, and décolletage creams.

But has this actually made our skincare any better?

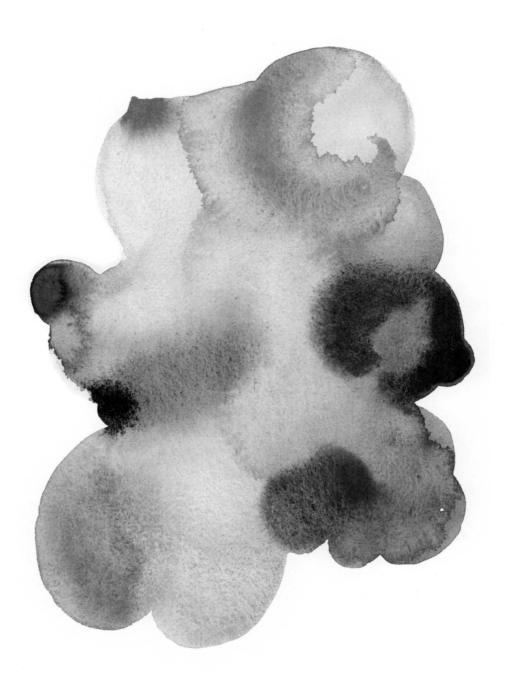

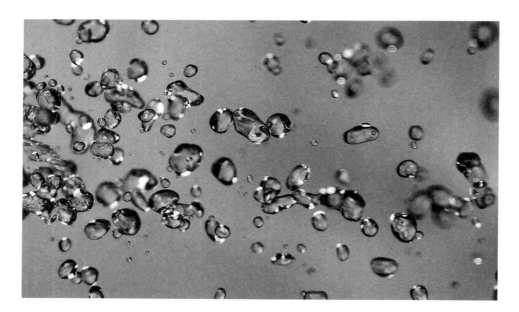

Less Is More

Though it's clear that not everyone needs skincare, those who do are often unhappy with the state of their skin, and they have to fight their way through masses of products to find one that suits them. Every item of packaging makes equally inflated claims, and every innovation is presented as a must, boosted by massive marketing campaigns. But more products don't mean better skincare. In fact, over-exposure to skincare products is a growing problem these days. There's no scientific evidence to suggest that serums, for instance, are any more effective than creams — and it's certainly not necessary to apply both serum and face cream.

Far from improving skin status, this plethora of products can actually make matters worse. The more products you use, the more preservatives, solvents, and other substances you're putting on your skin. So finding a skincare regime that works may mean going back to simpler solutions, fewer products, and a far lower skincare budget.

At the same time, different individuals have different skincare goals. Dealing with common skin problems like oily, dry, sensitive, or ageing skin is one thing. But contemporary notions of beauty go further, embracing the stereotype of a 'glowing' complexion.

FROM CLEANSING TO WRINKLES

Typically, a contemporary skincare regime begins with cleansing. The basic idea is that our pores have accumulated particles from urban air, dirt, grease, and other unhealthy substances, which need to be removed from the skin with a specific product.

However, scientific knowledge of the skin's barrier and its hydrolipidic film tells us that cleansing actually breaks down the skin's defences. It's no coincidence that there's a close link between cleansing and the most widespread skin problems, such as oily, dry, or sensitive skin and even rosacea.

So let's take a closer look at how we use cleansers, and how they affect the condition of our skin.

TENSIDES

Tensides are very widely used in soap and cleansing products, as they're highly effective at dissolving grease and removing dirt. They are surfactants, like the active ingredient in washing-up liquid. When scientists need to cause skin irritation for research purposes (to investigate anti-inflammatory substances, for instance), they apply a tenside, sodium laurate sulphate (SLS).

A recent Japanese study shows that tensides damage the hydrolipidic film. In other words, the more you wash, the more harm you'll do to the natural — and vital — sebum structures in the top layer of your skin.

Shower oils can also cause considerable irritation, as the tensides they contain remain on the skin, prolonging exposure. At one time I used to go to the gym with a friend called Sara. Out of the corner of my eye, I'd see white foam flowing onto the floor of the showers, and a glance would confirm that she was covered in it from head to toe. She'd hum as she rubbed even more sweet-smelling shower gel into her legs. Next afternoon, scratching her legs, she'd burst out: 'You're a specialist in skincare products — do you know anything that'll work for my legs? They get so terribly dry, and I haven't found a decent cream yet.' I advised her to take a close look at her shower gel and cleansers. The shower gel turned out to contain sodium laurate sulphate (SLS), which featured near the top of the ingredients list. She stopped using it straight away, and within a fortnight the itching was gone.

INCI: Sodium Lauryl Sulphate, Cocamidopropyl Betaine, Sodium Lauryl Sarcosinate, Tipa-laureth Sulphate, Coco Glucoside, Decyl Glucoside, Sodium Methyl Cocoyl Taurate, etc

Normal Skin

The ancient Greeks and Romans were already using antioxidant-rich oils on their skin.

People with normal skin have relatively little need of skincare. However, if you want to use a skin cream anyway, my advice is to look for one rich in antioxidants. Our skin has a lot to put up with — sun, exhaust fumes, and heat — all of which result in the formation of free radicals. What antioxidants do is break down these free radicals.

WHICH ANTIOXIDANTS SHOULD WE CHOOSE?

Most antioxidants are unstable and break down in a skincare product as it ages. A way to maximise stability is to select a package that keeps the antioxidant (vitamin C, for instance) separate, so you can add it in powder form. Or try looking for packaging with vacuum pump systems.

Looking beyond antioxidants, humectants such as glycerine, urea (carbamide), and N-acetyl glucosamine, and amino acids, can also be effective in hydrating the skin. This is particularly important in cold, dry climates. Niacinamide boosts the skin's barrier function and also has a sebum-regulating effect, reducing pore size and curbing inflammation.

Minerals such as magnesium, calcium, manganese, and selenium can also help to improve skin health. They have been proven to activate important enzymes in the skin and can boost the skin's natural antioxidant system.

The ingredients that are good for normal skin also provide a good basis for treating dry skin or oily skin.

SKINCARE THAT WORKS

ABOUT SKINCARE

In the Shower with Johanna

What's your skin type?

Apart from my vitiligo, my skin's normal in terms of skin status.

How does a skin scientist cleanse her skin?

When I take a shower, I don't wash with soap any more than necessary, and only under my arms. Everywhere else I just use water. To remove make-up in the evening, I use a mild cleansing cream. In the morning I don't wash at all, I just have a lukewarm shower.

Sensitive Skin

There hasn't been enough research into the reasons for sensitive skin, so it's hard to give general advice. However, it's worth thinking about why you have sensitive skin. Could it be linked with rosacea, for example?

If you have sensitive skin, it's more important to think about what you need to avoid than about what you should put on your skin. Avoid potential irritants such as alcohol, glycols, essential oils, other natural oils (in some cases), and preservatives such as methylisothiazolinone and other substances that release formaldehyde, phenoxyethanol, and benzoic acid. We're also more sensitive to preservatives and other substances after cleansing our skin — another reason to cut down.

Look for creams containing oils found naturally in the skin, and make sure your skin is exposed to as few substances that might cause a reaction as possible. Focus on anti-inflammatory substances tested for sensitive skin, such as allantoin, acetyl dipeptide-1 cetyl ester, or naringenin.

The same applies for rosacea-prone skin. A large study identified that the following factors worsen rosacea: too much facial cleansing, using facial masks more than four times a week, too-frequent visits to a beauty therapist, and use of make-up more than six times a week. Less is more!

INCI: Allantoin, Acetyl Dipeptide-1, Cetyl Ester, Naringenin

Dry Skin

Dry skin is mostly more sensitive than normal or oily skin. If you really must use a cleanser, choose a very mild cleansing cream.

After that, I'd recommend adding mainly substances that the skin produces naturally.

LACTIC ACID

Urine and lactic acid have both been used in skincare since ancient times. Two thousand years ago, Cleopatra bathed in sour milk — which contains high levels of lactic acid.

Use creams containing plenty of fats found naturally in the body, such as squalene and ceramide. You should also add humectants — examples are glycerine, urea, lactic acid, amino acids, or N-acetyl glucosamine.

Oily Skin

If you need to curb oil secretion in your skin, there are effective skincare products available — plus substances you should avoid, and others you should add.

A common misconception about people with oily skin or problem pimples is that they need to wash their skin frequently with strong cleansers. In 2018, a team of dermatologists from NYU's School of Medicine published an article analysing fourteen studies of acne sufferers and cleansers, based on data from a total of 671 informants. The results showed that cleansing doesn't improve skin with acne.

Certain waxes, butters, fatty alcohols, fatty esters, and oils can be comedogenic, meaning they tend to *cause* break outs. The list of substances to avoid includes lanolin, certain fatty alcohols like myristyl myristate, fatty esters, and butters such as cocoa butter. However, that doesn't mean all waxes, butters, fatty esters, and oils are comedogenic. Moreover, these substances affect different individuals differently, and you need to look at a product as a whole.

Oily skin can often benefit from applying acids, which reduce pore size. Zinc is another ingredient that can help people with oily skin.

This is by no means a new discovery, as zinc has been used to treat greasy skin and acne since the 1930s. Skincare products containing retinol and niacinamide have also proven to be effective in treating oily skin.

THREE PRODUCTS TO TREAT OILY SKIN

Linda, a friend of a friend, got in touch with me after giving birth to her first child:

'I've had a greasy complexion all my life,' she said. 'When I was a teenager, my greasy skin broke out in pimples. And though I don't get many of those these days, I've still got oily skin. I'm pushing forty now, I'm a mother, and I still have this problem. What can I do about it?'

As we've seen, there's nothing wrong with oily skin as such. But if you want to reduce the oiliness of your skin anyway, here are a few tips that should help.

I started by advising Linda to wash as little as possible, and only in the evening, using a mild cleanser to remove make-up.

Then I formulated two products for her. They contained no preservatives, emulsifiers, waxes, butters, or silicones, and were formulated as fairly light oil gels.

I formulated a day cream containing no UV screen. Instead, I advised Linda to use a separate sunscreen if she spent any time out of doors in strong sunlight. The cream had a light, gel-like consistency, formulated with squalane. I added a specific plant stem cell extract that has both anti-inflammatory and antioxidant properties, plus niacinamide for its sebum-regulating and barrier-strengthening qualities.

The night cream was also based on a fairly light gel emulsion, to which I added zinc, retinol, and a fairly mild acid (PHA). I advised Linda to keep these products in the fridge, as they contained no preservatives. Moreover, substances such as retinol and plant stem cell extract are temperature-sensitive.

After two months, Linda felt her skin was much better, and she'd stopped using foundation altogether, for the first time since she was sixteen.

Skin That's No Longer Twenty Years Old

The last two decades have seen an explosion of anti-age products, which now dominate the skincare market. They take their cue from plastic surgery, and the accompanying sales pitch typically claims they can give you a facelift, reduce wrinkles, and eradicate discoloured patches. It's also clear that the anti-ageing market is where skincare companies are investing their money.

But do these products have the same effect as plastic surgery, a botox injection, or laser treatment? No, of course they don't.

Clinical studies of the impact of different types of skincare treatment show results ranging from a few per cent to a 60 per cent reduction in wrinkles. Some ingredients work, others don't. Retinol remains the most studied rejuvenating substance, and it's been shown to reduce wrinkles, lighten skin, and reduce pigmentation. Interestingly, it also makes the epidermis thicker.

There are substances that have a rejuvenating effect, but they can also irritate the skin and make it red if you use too high a concentration. This applies particularly to people with sensitive skin.

My recommendation is to protect your skin against ageing too rapidly by using creams containing antioxidants, protecting yourself against direct sunlight, and applying creams with ingredients that

inhibit the MMP enzyme. Combining retinol with an MMP-inhibiting antioxidant can be an effective strategy.

As we age, there is an increase in local skin pigmentation, resulting in patches of pigment. These changes can be treated to some extent with acids. As we've seen, acids speed up the transfer of melanin from the basal layer of the epidermis to the outermost skin layer. However, if the acid

also irritates the skin, the opposite effect can result, leading to hyper-pigmentation — precisely what the treatment was intended to reduce in the first place.

However pretentious this may sound, it would be great if we could escape this anti-age hysteria and instead use creams that are good for our skin — ones that'll help it stay healthy. The bonus might even be fewer wrinkles.

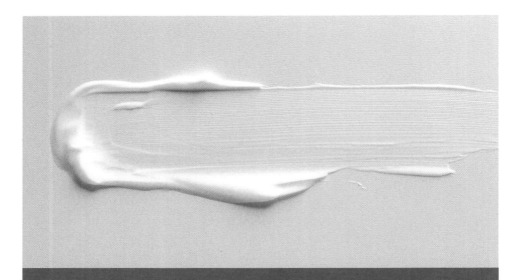

Pick the Right Product

Look for a cream with as few ingredients as possible, and beware of products consisting mainly of emulsifiers, preservatives, or masking agents.

Then look for a humectant such as glycerine, urea, amino acids, or N-acetyl glucosamine, a building block of hyaluronic acid.

As regards emollients, it's best to go for oils that occur naturally in the human body, such as squalene, ceramides, and cholesterol.

The same applies to active ingredients. We can give our skin a boost by adding substances already found in the skin but which dwindle as we age or when we're exposed to UV radiation, such as amino acids, minerals, and vitamins.

Finally, go through the list of ingredients to identify any acids. How the skin reacts to acids varies enormously from one individual to another. Remember to be cautious with acids during the warmest six

months of the year, as there's a risk of thinning the skin, which may result in irritation or, in the worst case, more pigment patches. If you want to use acids to make your skin more even, try out products containing PHA first, then ones containing AHA / BHA if your skin has pigment patches or enlarged pores or is oily.

If a cream is marketed as containing an attractive substance (caviar, diamond powder, or mineral water, for example) that supposedly has some effect on wrinkles or pigment patches, or that is supposed to be particularly good for hydrating the skin, it shouldn't also contain substances that are clinically proven to have the desired effect, such as retinol, vitamin C, niacinamide, or other substances, such as acids. You should know where the effect comes from.

What Do Wrinkles Signify?

No face cream on this earth can make wrinkles disappear altogether or give you a facelift. Instead, it's best to focus on creams that give you healthy skin, reduce collagen degradation, and have a preventive effect. If, for whatever reason, you can't live with wrinkles, you'll need to use other methods — botox or plastic surgery.

A few years ago, my team and I decided to look at what aspects of a face really affect observers' perceptions, by conducting an age perception study. This study, held in Russia, involved using a clinical camera to take photos of Russian women aged forty to sixty-five, after which 200 Russian women looked at the pictures and guessed the participants' age. We then compared the results with a variety of skin parameters, including wrinkles, irregular pigmentation, redness, and loose skin. While wrinkles made people look older, it was interesting that the result depended mainly on where they were. Lines on the forehead or under the eyes were significant, as was the nasolabial fold (running from the nose to the corner of the mouth), while crow's feet around the eyes were less important. This is particularly interesting, given that crow's feet are often linked with laughter lines and a happy expression.

And yet many anti-ageing treatments focus precisely on crow's feet. Irregular skin tone — changes in pigmentation or red areas — also made women appear older than they were. We published the study in 2015 in the *International Journal of Cosmetic Science*.

Another interesting study looked at whether facial redness affected perceptions of attractiveness. It was found that an even reddish skin tone was considered more attractive. The colour is associated with higher levels of blood oxygen resulting from physical exercise or a sauna.

To sum up, the ideal has shifted from the white-painted faces of the eighteenth century to the tan of the 1980s, and now to today's flushed faces, which suggest long hours spent in the gym or out jogging. I think anti-ageing products will eventually be phased out of the market. A number of trend reports suggest this, and many skincare companies have taken it up as a strategic plan. Instead of *anti-ageing*, we're going to see expressions like *pro-ageing* and *well-ageing* on skincare packaging.

SKINCARE THAT WORKS

Marie Lodén
Pharmacist, Lecturer, and Skincare Expert

We meet in Marie's laboratory in Solna, relax in armchairs, and sip sugar-free raspberry cordial from wine glasses. Marie and I have met many times over the last few years, sometimes at her courses, where I've appeared as a guest lecturer, and sometimes at other research-related events. Marie holds a doctorate on the impact skincare products have on the skin, and she's been a head of research and development for many years.

ABOUT SKINCARE

Which humectants do you rate most highly?

Definitely carbamide. I think it's better than glycerine. But it needs to make up at least 5 per cent of the product in order to reduce evaporation and hydrate the skin effectively.

Can the skin get used to skin creams and actually become drier in the long run?

Yes, there's research to show that happens. We see that with creams that don't contain any humectants, such as glycerine or carbamide, just fats like emollient oils or waxes. They ruin the skin. Evaporation increases, and the barrier function is weakened.

What are your views on so-called natural skincare?

It sounds marvellous, doesn't it, but there's no particular advantage in skincare being 'natural'. There's so much in nature that's actually harmful to the skin. We humans are natural as well, and we produce many substances ourselves, like fatty acids, ceramides, and oils. Snail secretion is the big thing right now, but why should something that comes from a snail or from dog roses be any better than substances that come from humans?

What would you say to someone who wants to buy a day cream with UV screen in autumn?

UV screen stops the skin making vitamin D, so it's a big mistake to wear it all year round. Many UV filters aren't good either for the skin or for the environment, so it's a good idea to do without them when you can. But when you're on holiday in a sunny place you absolutely must use a UV screen. I belong to the generation who used tinfoil funnels and sun lamps in winter, so now I've got a lot of discoloured patches on my face. If I had my life over again, I'd definitely be more careful. I wouldn't necessarily have used sunscreen, but I wouldn't have sun-bathed so much, and I'd have protected myself in other ways, by wearing suitable clothing and staying in the shade.

There's no particular advantage in 'natural' skincare.

Eye contour creams — what should we bear in mind?

It might be useful to remember that creams tend to creep along furrows in the skin. When you apply an eye contour cream, you may think you've kept it well away from your eye, but it can still cause a reaction.

Mineral oil, paraffin, Vaseline — there are many names for what's essentially the same thing. What are your views on these products?

There's nothing wrong with them. They're non-reactive and don't harm the skin. You don't see any reactions from mineral oils and you don't get blocked pores — which does happen with plant-based oils.

Have you observed any differences between mineral oils and natural oils?

No — they have the same emollient effect. We compared mineral oil with rapeseed oil in one study and the results were identical.

Is there any super-ingredient you'd recommend?

Yes — it's been shown that retinol has an effect on wrinkles and that it makes skin firmer.

What do you use yourself?

I use creams from time to time when my hands feel dry, and I sometimes use a tinted face cream to even out my skin tone. But I don't use any other kind of skincare. I've never felt the need for it.

Thank you, Marie!

A Plethora of Products

We're in the skincare section of a big department store. A sales assistant with a glint in her eye swoops down on a forty-something woman, pinching the skin under one eye.

'Your skin's awfully thin just here,' she says. 'The delicate skin under your eyes needs a lot of extra pampering, you know. And they've developed special creams just for this specific type of skin.'

She turns to take down a few little packages from the display case shelves. The woman nods, looking scared. A few minutes later, she leaves with an expensive 15-millilitre tube. Fifteen millilitres corresponds to a single tablespoonful of the product.

Eye Contour Cream

If you have bags under your eyes or dark circles around them, an eye contour cream might well be helpful. Bags under the eyes are often the result of a small amount of fat and liquid accumulating there. If you don't want to resort to plastic surgery, you can break down the fat with the help of a substance such as caffeine. This is used in nearly all eye contour creams, as well as in anti-cellulite creams.

Serum

'Invest in a magical elixir that will enhance your skin at the very deepest level.'

This is the kind of claim you might find on a vial of serum. But as usual, if something sounds too good to be true, there's a catch — it isn't!

So how exactly does a serum differ from a day or night cream? Well, the biggest difference is the consistency. Serum is usually a less-viscous gel or oil gel. There are no regulations on what ingredients it has to contain. Serums are often marketed as highly effective products with high levels of active substances and better penetration capacity than ordinary day and night creams. But that's not necessarily true.

Certain substances do make it easier for a product to penetrate the skin. Organic solvents such as glycols are an example — but these substances can just as well be added to a cream.

There are absolutely no studies to show that a serum is any more effective as an antidote to wrinkles than a cream. What matters in each case are the ingredients.

Day and Night Creams

The biggest difference between day and night creams is that the former contain a UV filter and are often lighter in consistency than night creams.

Manufacturers may also avoid using certain substances in day creams and add them to night creams instead. An example is retinol, an active ingredient which can make the skin more light-sensitive.

A night cream is a little creamier in consistency than a day cream, especially if it's intended for normal or dry skin. Night creams often contain active ingredients such as retinol and acids.

Ask Johanna:

Don't eye contour creams help remove wrinkles around the eyes?

If you have wrinkles around your eyes and want to use an anti-wrinkle cream, it's best to invest in one that you can apply to the rest of your face as well. You probably want to boost the production of collagen and elastin in other parts of your face, after all. There's no magic formula that only works around your eyes.

However, if you have specific issues in your under-eye area that you don't have in the rest of your face — such as puffiness or dark circles — then it might be good to invest in a specific eye product.

Face Masks

There are no scientific studies to show that face masks are any more effective than day or night creams. Nor are there any rules on what they can contain or on the concentrations of individual ingredients.

Face masks are often marketed as a way to give the skin an extra boost; this can mean providing extra hydration, making the skin smoother, or reducing pore size, for example. One advantage they have is that the demands placed on them by the consumer are different from those placed on face creams. The main reason for this is the length of time users leave a mask on for before removing it. Since it only stays on your skin for a limited time, the ingredients can be so concentrated that the consistency would be too sticky in a day or night cream. Treatments can also be formulated to be applied for a shorter length of time — for instance, with higher acid levels.

Peels

There are endless different peeling products and facial scrubs on the market. Their function is to remove dead skin cells from the face, using everything from crushed apricot kernels that slough off dead skin to exfoliating acids that thin the epidermis. A term we're hearing more and more is 'enzyme peel'. This isn't anything new, nor does it have anything to do with enzymes. What it actually contains is an exfoliating acid.

I have nothing against peeling products, provided they contain no environmentally hazardous substances such as microplastics. Microplastic particles are washed down the plughole and into our streams, rivers and lakes, where they are eaten by fish and

subsequently by animals further up the food chain. What's dangerous about them is that they can attract environmental toxins, and they aren't broken down by the digestive system. Microplastics on the INCI list have names such as polystyrene/acrylate copolymer, nylon, polyethylene (PE), or polymethyl methacrylate (PMMA).

Many countries, including Sweden, the US, and the UK, have banned the use of microplastics in rinse-off products.

Provided that your peel is free of environmentally harmful substances, it can provide a useful way to remove dead skin cells from the stratum corneum once a week or so.

Make-Up and Foundation

Make-up and foundation can clog the pores, which naturally need to be able to breathe — and the results are bad for acne sufferers. An interesting study conducted in Poland in 2017 showed that people who used full coverage foundation made their acne worse. This can often lead to a vicious circle. The more you cover up your skin, the worse it gets.

Johanna on Facial Toners

Toners are marketed with claims that they provide benefits like reducing pore size, hydrating the skin, and removing traces of cleansing product. But I just can't see the point of this product. Here are my comments on seven claims made for toners in a beauty magazine that came out in 2018:

They makes pores look smaller

For that to be true, a toner would have to contain an ingredient for that specific purpose. An acid, for example — which could equally well be added to a day or night cream.

Their pH value is better for the skin than the pH of tap water

Many skincare manufacturers would have us believe that water is terrible for the skin because of its pH, which lies between 7 and just over 8, making it slightly basic. But there's absolutely nothing wrong with putting tap water on your skin. Remember, for nine months before we're born, we float in amniotic liquid, which is also slightly basic. It's hard to live without water on your skin.

They hydrate the skin

It's true that a toner often contains a humectant, such as glycerine — but so do any creams you use, so one extra layer isn't going to make any difference.

They remove traces of cleanser

If you need to remove traces of cleanser which haven't been removed by lukewarm water, you're not using the right cleanser.

They help your serum, eye contour cream, and moisturising cream penetrate the skin better

The only way to do this would be to use a solvent on your skin. Many glycols, for example, are skin-penetration enhancers — but I wouldn't recommend using them if you want your skin to remain healthy in the long term.

They protect you against air pollution

There are two ways in which this could be achieved. One would be to apply a polymer that covers the skin with a film — but it wouldn't be too smart to do that just before applying your day or night cream. The other would be to use a toner containing special substances that give proven protection against the free radicals generated by environmental pollution. Such substances exist, but you might just as well add them to a day cream instead.

All in all — toner is a totally redundant product!

Johanna Introduces ...

Lars Norlén
Dermatologist at Karolinska University Hospital's Skin Clinic and at Sturebadet Health Care

Lars and I met for the first time ten years ago at a conference in Copenhagen. Lars and his team are global leaders in research into the skin's barrier function. As a doctor, he sees many patients seeking medical advice.

ABOUT SKINCARE

What would you say are the most important ingredients in skincare products?

Humectants — no doubt about it! I recommend glycerine as a substitute for water in the skin. Many skin creams designed to prevent eczema or extremely dry skin have high glycerine levels of between 10 and 20 per cent. That gives them effective humectant properties.

As one of the world's leading experts in the barrier function, would you say it's possible to hydrate the skin by drinking water?

Let's be quite clear — that's just nonsense. The more water we drink, the more we lose by passing water. What does work are the humectants in skin creams, which bind water in the skin and replace water in dehydrated skin.

What cleanser would you recommend?

As a medical dermatologist, cleansing isn't something I'm concerned with. Since there are no studies so far that show a clear improvement in skin status thanks to cleansing, we don't specifically recommend it.

What about acne — what treatment do you recommend?

I always recommend treating the skin directly, rather than taking tablets. There's a lot you can achieve provided that the patient follows the treatment to the letter. As far as possible, it's best to avoid taking antibiotics, which can lead to drug resistance and other side-effects. If there's a risk of scarring, though, it may be worth considering antibiotics.

At your clinic, do you see fresh outbreaks of acne after treatment with antibiotics?

Yes, we do.

Drinking water to hydrate your skin is nonsense.

What are your views on cosmetic procedures such as acid treatments?

Chemical peels tighten the skin and make it slightly smoother. But that's not necessarily good for the skin as such. My overall opinion is that if having slightly smoother skin makes you feel better, that's fine as long as it doesn't harm the skin. But you shouldn't use medical treatments on healthy skin unless absolutely necessary.

What's the most interesting area of research at the moment?

Research into the barrier function is advancing at an amazing rate. We now know how the skin barrier is organised — right down to the atomic level. The more we understand about the structure of the skin, the more we can learn about how to develop new molecules and treatments.

Thank you, Lars!

Sunscreens

Apart from clothes, a hat, and a parasol, sunscreen is the best protection there is against the sun. It's important to use it, both to avoid sunburn and because the sun is by far the main cause of wrinkles, sagging skin, discoloured patches, and uneven skin tone. However, UV screen has no beneficial effects on the skin apart from protecting it against sunlight. That's why we should avoid using it when we're not out in the sun. It's great to wear a seatbelt when you're driving, but quite unnecessary (and pretty uncomfortable) when you're on the sofa at home.

Fortunately, the glory days of the 1980s are over. There's no tinfoil as far as the eye can see. And in many countries — especially those with a few more hours of sunlight than Scandinavia — people now avoid the sun like the plague.

The UV Index

The UV index is an international standard used to measure how intense — and thus harmful —UV rays are. Using the UV index, we can compare how intense UV rays are at different places and times. The scale is linear, and the higher the index value, the stronger the UV radiation. An illustration of the UV index in Figure 1 shows how the UV index usually looks in Sweden).

In Scandinavian countries, the UV index is usually between 4 and 7 during the summer season, and lower than or equal to 2 during the winter, which may explain our longing for the sun each spring! At low UV index, no sunscreen is required, but at high UV the risks of negative effects of the radiation increase, and you need to protect yourself.

The UV index varies mainly by season and time of day. Other factors are the weather and the thickness of the ozone layer. Around the Mediterranean, the index is high or very high — between 7 and 10 — and extremely high levels (over 10) can occur at any time of year in equatorial zones.

In Australia in January, for example, the UV index is extremely high: 11+ in almost the entire country. In the US, the UV index varies between regions, with increased UV exposure occurring closer to the equator. In the UK, the UV index does not exceed 8 (and 8 is rare; 7 may occur on exceptional days, mostly in the 2 weeks around the summer solstice).

In the northern hemisphere, the sun reaches its zenith at the summer solstice in June. This is when we are exposed to the highest levels of UV radiation. The opposite is true, of course, for the southern hemisphere: the summer solstice there occurs in December.

During winter, the sun is so low in the sky that UV radiation is at its minimal point. In late winter, however, in countries where it snows, the UV index can virtually double if the ground is covered in snow, as

INDEX

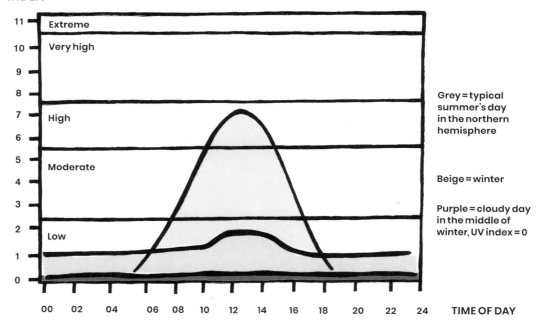

Grey = typical summer's day in the northern hemisphere

Beige = winter

Purple = cloudy day in the middle of winter, UV index = 0

TIME OF DAY

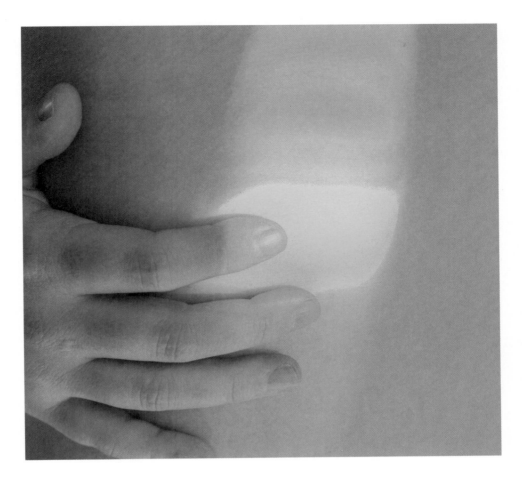

snow reflects the sun's rays very effectively.

During the six warmest months of the year, I recommend applying sunscreen whenever you're out of doors, whether just for a short while or all day, no matter where you are. Use a high factor, preferably 50. You should also make sure the cream protects against ultraviolet light of different wavelengths, both UVA and UVB. It is important to protect against both UVA and UVB radiation. For products marked with the UVA symbol, the UVA protection is at least one third of the SPF value with which the product is labelled. These properties will be detailed on the packaging. UVA rays penetrate deepest into unprotected skin, right down to the dermis, while UVB rays are reflected off the epidermis. Sunscreen reflects these damaging rays back before they even reach the skin.

In Australia and in countries where UV index is always above 2, during the day, it is very important to protect against the sun using protective clothing, sunglasses, hat, and sunscreen.

In countries where the UV index is below 2 during the winter season (for example, the UK), sunscreen is not necessary.

Chemical and Physical Filters

The different types of filter are what protect you in a sun cream. We call them chemical and physical filters, or organic and mineral filters.

For a product to provide a high sun-protection factor that is suitable to protect against both UVA and UVB, it needs to have both chemical and physical filters. Many manufacturers use more chemical filters, as these products sell better. This is because they don't leave a sticky white layer on the skin. Products aimed at children, on the other hand, go easy on chemical filters, though they still combine both chemical and physical aspects.

The chemical filters are the most numerous. For them to be effective, they need to be dissolved in oils. Some chemical filters are broken down by sunlight: they are 'photo-unstable', which reduces their sun-protection properties. In some cases, this can also lead to the occurrence of harmful substances that cause 'photo-allergies'.

Chemical filters only have the effect of UV protection when dissolved in other substances in the sunscreen product, such as oils and waxes. The most skin-friendly and effective UV filters have both a high molecular weight and a high melting point, which means that they do not penetrate the skin, but it requires a lot of oil to dissolve them in the formulation. This can make the product feel sticky to apply, especially on a hot and sweaty day.

Now we will deep-dive into the chemical filters. I have tried to simplify, but the situation is complex. It is difficult for most consumers to understand which filter is in the product, as most chemical filters have several names. In addition, the use of filters varies in different parts of the world. Several of the more newly developed chemical UV filters are restricted for use by the FDA and that means that the US market uses many filters that have been phased out in Europe and most parts of the world, due to safety risks.

OXYBENZONE — ONE OF THE MOST WORRISOME UV FILTERS

One of the UV chemical filters that gives most cause for concern is oxybenzone. Oxybenzone is the filter that penetrates deepest into the skin. From the chapter about the active ingredients in skincare, you'll recall that 500 daltons is a threshold value for skin penetrability. The oxybenzone molecule measures 228 daltons, so it can readily make its way through the skin.

An American study showed that Oxybenzone was detected in 96.8 per cent of urine samples from over 2500 people tested. Once it's penetrated the skin, it goes right through our system — yet the whole idea of this filter is that it's supposed to cover our skin and absorb UV radiation.

In a study published over twenty years ago, Professor Karin Schallreuter showed that oxybenzone is responsible for the inactivation of important antioxidant systems in the skin.

Although there are sixty-seven scientific articles on this filter's negative effects, it is freely used worldwide in concentrations of up to 6 per cent.

UVB Blockers

AMINOBENZOATES

Aminobenzoates were one of the first UV filters. They are the most potent UVB absorber but do not absorb UVA. Their use has declined due to para-aminobenzoic acid (PABA) sensitivity. PABA is a very effective UVB filter; however, it was reportedly the most common photo-allergen and contact allergen, and is now phased out.

CINNAMATES

Cinnamates have replaced PABA as the next-most potent UVB absorbers, and include octinoxate (ethylhexyl methoxycinnamate) and cinoxate. Octinoxate is the most commonly used UVB filter in the United States. However, octinaxate is not very photo-stable and degrades in the presence of sunlight after a short period of time. Cinnoxate is less commonly used.

SALICYLATES

Salicylates are weak UVB absorbers and therefore used at higher concentrations. They are also used to increase the effect of other UVB filters. Examples of these are homosalate (trimethylcyclohexyl salicylate) and octisalate (ethylhexyl salicylate.

Octocrylene is a very commonly used chemical filter but has recently also been associated with photo-toxicity and photo-allergic potential.

ENSULIZOLE

Ensulizole is a pure UVB filter. It is a water-soluble compound that is commonly used in skincare products to give a lighter, less oily feel. Although reported to cause photo-allergy, it is considered to be a low level sensitiser.

UVA Blockers

BENZOPHENONES

The most common trade names are Benzophenone-3 and Oxybenzone. Benzophenones absorb mostly UVB, while oxybenzone is considered a broad-spectrum absorber as it can absorb UVA as well. However, of all sunscreens, oxybenzone has the greatest likelihood of inducing contact or photo-contact dermatitis.

AVOBENZONES

Avobenzone is a dibenzoylmethan derivative that has been used in sunscreen since the 1980s. Able to absorb the full spectrum of UVA, it doesn't, however, provide protection against UVB. Its drawback is that it's very photo-unstable, and can lead to contact dermatitis.

ECAMSULE AND DROMETRIZOLE TRISILOXANE

Ecamsule is made up of terephthalyidene dicamphor sulfonic acid, a very photo-stable product, and seems to stay on top of the skin with limited risk of penetration, as does drometrizole trisiloxane. Ecamsule is approved by the FDA. Drometrizole trisiloxane is not yet approved by the FDA. Approved in EU, Canada, Australia, and Japan.

BROAD-SPECTRUM SUNSCREENS

Methylene-bis-benzotriazolyl tetramethylbutylphenol (MBBT) (Tinosorb M) and bis-ethylhexyloxyphenol methoxyphenyl triazine (BEMT) (Tinosorb S) are examples of broad-spectrum filters. Both filters are large molecules, which decreases the likelihood of systemic absorption or endocrine effects. They are also considered to be photo-stable and non-irritating to the skin.

None of these filters are approved by the

SUNSCREENS

FDA, though, and are therefore not available in the US. They are available in Europe, Australia, and other countries.

Chemical sunscreen (INCI): Benzophenone-3 (Oxybenzone), Octylmethoxycinnamate / Ethylhexyl Methoxycinnamate (Octinoxate), Homosalate (Trimethylcyclohexyl Salicylate), Octyl salicylate/ Octisalate (Ethylhexyl salicylate), Octocrylene (2-ethylhexyl 2-cyano-3,3-diphenylacrylate), Butyl Methoxydibenzoylmethane (Avobenzone), Bis-ethylhexyoxyphenol Methoxyphenyl Triazine (Tinosorb S), Drometrizole Trisiloxane (Mexoryl XL), Methylene-bis-benzotriazolyl Tetramethylbutylphenol (Tinosorb M), Ecamsule (Mexoryl SX).

Physical Filters

The mechanism of physical sunscreen is based on the reflection and scattering of UV light, pretty much in the same way as clothing. Physical filters are the ones that cover the skin in a white layer. Zinc oxide and titanium dioxide are often combined to provide optimum UV protection. Zinc oxide protects against a wide range of UVA, is very photo-stable, and doesn't react with other UV filters. It is more effective than titanium dioxide for UVA protection; however, it is less efficient against UVB radiation.

Titanium dioxide protects against UVA and UVB, but it doesn't protect against UVA (340 to 400 nm); like zinc oxide, it is photo-stable and doesn't react with other filters.

Essentially, physical filters do not penetrate the skin but remain on the surface of the skin, where they can protect against harmful radiation. Since physical filters cannot penetrate the skin, the body is less likely to react to them than to chemical filters. It is, among other things, the low risk of allergies and the lack of hormone-disrupting effects that mean physical filters

are often used for products for children and for those with sensitive skin.

Many consumers dislike the white coating that physical filters leave on the body. As a result, manufacturers have started to produce filters with smaller particles, known as nanoparticles. The smaller the particles, the less of a white layer the filter leaves on the skin. Nanoparticles are too large to be able to penetrate the skin, but they can be dangerous if inhaled. Moreover, they can be toxic to corals, fish, and other organisms.

Since 2013, it has been mandatory in the EU to label products containing nanoparticles. For instance, a sunscreen might be labelled as follows: Zinc Oxide (Nano). The EU took the lead on nanomaterials, and, in truth, allowed them in cosmetics too early, as much was unclear about their use, like how to measure their size. There is no requirement to label nanomaterials in the US or Australia.

If concerned about ingestion of nanoparticles, my advice would be to avoid sunscreen aerosols and sprays, where the risk is higher of inhaling them, and use lotions or creams instead.

Physical sunscreen (INCI): Titanium Dioxide, Zinc Oxide

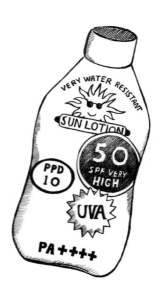

Environmental Aspects

All sun protection products are resource intensive and have an environmental impact, which comes from producing the raw materials as well as the product itself, its packaging, and all associated transport. These products also affect the environment after we use them. Aquatic organisms and their ecosystems are particularly affected, because most of the products we use eventually end up in the water — when we swim, when we shower after the beach. or when we wash clothes and towels with residual sunscreen. It's estimated that 25 per cent of the sunscreen ingredients we apply end up in the water. A number of scientific publications have attempted to map the effects of various UV filters on the environment. Certain chemicals — including oxybenzone, octinoxate, octocrylene, octisalate, avobenzone, and homosalate — have been identified as particularly dangerous for ecosystems, making coral more susceptible to bleaching, deforming baby coral, and degrading its resilience to climate change.

The United States Virgin Islands recently enacted a ban on the importation, sale, and possession of sunscreen products containing the active ingredients oxybenzone, octinoxate, and octocrylene. Hawaii and Key West, Florida, are poised to ban oxybenzone and octinoxate being used in sunscreens. In the rest of the US, however, both filters are permitted (oxybenzone at a maximum level of 6 per cent, same as in the EU).

The more skin-friendly physical filters, titanium dioxide and zinc oxide, have also been shown to adversely affect aquatic organisms such as fish, algae, and crustaceans.

The situation is complex, but in order to protect adequately against the sun, my advice will be to use a sunscreen that contains safe chemical filters and physical filters, but not to rely on sunscreen alone. Avoid the strongest sun during the day, and wear protective clothing, hat, and sunglasses.

SUNSCREENS

Dress to Protect!

Clothing is the first line of defence against the sun. The right clothes can offer more sun protection than sunscreen, since people often don't use sunscreen in the right way (unevenly applied and forget to reapply).

How do different items of clothing affect how well you're protected? How can you be sure a piece effectively blocks the sun's rays?

Here are some tips:

Colour

Dark clothing — black and navy-blue — absorbs more UV rays than lighter colours, like whites and pastels. For example, a white cotton T-shirt has an SPF of only 10. As a rule of thumb, the darker the hue, the better protection the clothing will provide.

Material

Like colour, the material, weave, and texture of your clothing can affect how well it protects you from UV rays. Synthetic and semisynthetic fibres like polyester or rayon are the best choices for sun protection, as are dense, heavy, tightly woven fabrics like wool, denim, or corduroy. On the opposite end of the spectrum are lightweight fabrics (for example, refined cotton), which tend to be thinner and let more light pass through.

- Denim shirt — SPF 1700
- Viscose blouse — SPF 15
- Cotton T-shirt — SPF 10

Size

It's pretty obvious that the more skin you cover, the better protected you are. It can be easy to forget that the same thing applies to hats! The best hats for sun protection have a wide brim (7–8 centimetres or greater). Look for a tightly woven hat, rather than a straw hat that allows UV to pass through its openings. Don't forget to check the fit of your sunglasses too — a pair that slips down your nose is leaving your eyes at risk of sun damage. Look for sturdy sunglasses with wide lenses that cover the eyes, eyelids, and as much of the surrounding areas as possible.

Fit

A loose-fitting shirt provides better SPF than a tight one.

Ask Johanna:

What sunscreen do you use?

I use a sunscreen developed for children, which has a protection factor of 50. I use both chemical and physical filters, but the chemical filter has low skin penetration.

How careful are you to apply sunscreen?

I apply sunscreen to my face every day during the six warmest months of the year, and I always use it when I'm on holiday in a sunny place or on a skiing holiday. Snow reflects the sun's rays.

How about on a cloudy summer's day?

If it's a cloudy day in summer, I apply factor 50 anyway. You're out of doors much more often in summer, and the weather can change quickly. There's a big risk that your skin may be affected.

And how about winter?

I don't use any UV protection in winter, not even in my day cream. It's quite unnecessary in November, for example, when I'm mostly indoors, or at the most walking round Stockholm for a few hours when the UV index is low.

What if you moved to Australia or Florida?

That is very tempting some days, I must say! I would avoid the strongest sun during the day (from 10am to 4pm), wear protective clothing, use sunscreen — SPF 50, at least — on body-parts exposed to the sun, invest in a cool hat, which provides shade for my face, and wear sunglasses.

About the Micro-biome

New discoveries have shown that micro-organisms play a bigger role in our skin health than we could ever have imagined. We're at the start of a bacterial health revolution.

CHAPTER 9: The bacterial revolution
CHAPTER 10: Bacteria and fungi in the skin
CHAPTER 11: Boost your microbiome

The Bacterial Revolution

Over the last few decades, research into the microbiome has boomed. Every one of us hosts countless trillions of micro-organisms belonging to thousands of different species: bacteria, viruses, and fungi. There are so many of them that we actually have more bacterial genes than human genes in our bodies! Our micro-organisms live in our large intestine, nose, and mouth — and in our skin as well.

Thanks to the bacterial revolution, we've started to cooperate with the micro-organisms within us, instead of fighting against them. Perhaps you're already part of the revolution? Maybe you've experimented with fermenting vegetables, or perhaps you make sure to include prebiotic fibre in your breakfast?

It's time to get to know the skin's very own microbiome.

ABOUT THE MICROBIOME

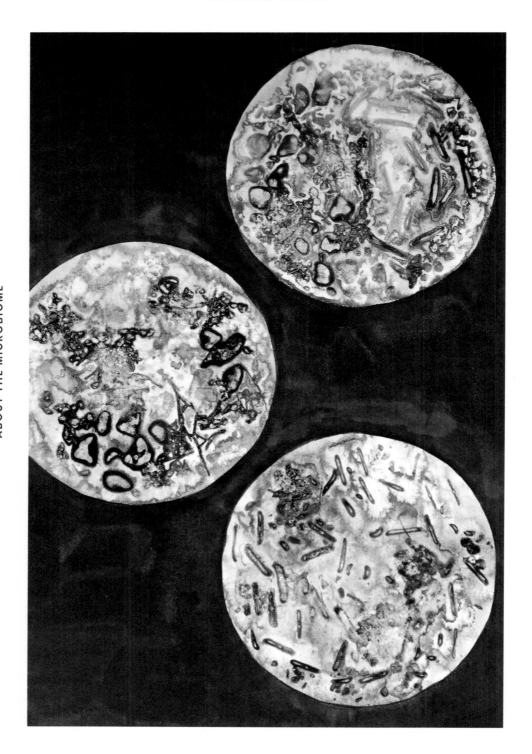

Micro-Organisms Make a Comeback

Bacteria have long had a bad press. There's a certain irony in that — after all, they were here on earth long before we were. They colonised this planet around three billion years ago. Compare that with the history of modern humans, which began a mere 300,000 years ago.

Without micro-organisms, we humans couldn't survive. They cooperate with our cells and play an essential role in all aspects of our health. In fact, they control most of what goes on, including our daily lives. Do you think you've got a weak character if you can't resist eating sweets in the evening, or that you're unlucky if you constantly feel off-colour?

That's not the way it is.

Micro-organisms regulate your sweet tooth, they break down the food in your digestive tract, they protect you against the unwelcome microbial guests you encounter daily, they cooperate with your immune system — and they even interact with your brain and your personality.

With these new insights, today's scientists are developing a new perspective on micro-organisms, reclaiming a balance between bacteria, viruses, and fungi as the key to good health.

HUMANS DISCOVER BACTERIA

The Dutch scientist Antonie van Leeuwenhoek first observed bacteria in 1674. He saw the tiny organisms through a microscope that he had built himself.

In 1861, the French chemist and biologist Louis Pasteur proved that fermentation is the result of microbiological activity, and it became known that bacteria can cause illnesses. In 1928, the British bacteriologist Alexander Fleming discovered penicillin, apparently giving humanity control over invisible colonies of bacteria that posed a constant threat. Yet even then Fleming warned about the side-effects of antibiotics.

Today, ninety years later, we're facing a post-antibiotic era, with antibiotic resistance and other illnesses as a result. Antibiotic treatment has a major impact on our gut microbiome, and it can take months for the body's microbiome to recover fully. Side-effects such as an upset stomach or diarrhoea are likely results of antibiotic treatment, as are rashes and fungal infections of the skin.

THE BACTERIAL REVOLUTION

The Gut–Skin Axis

There's much to be said for Hippocrates' dictum — now nearly 2500 years old — that every-thing begins in the gut! Because it really does. The bacteria in your gut are a precondition for a healthy skin. An imbalance in your intestines can lead to skin disorders such as acne, eczema, and psoriasis.

Put together, your gut bacteria weigh about 1.5 kilos. That's a lot of bacteria — as many as forty billion of them, according to recent studies. These masses of bacteria break up nutrients that reach the large intestine, producing short fatty acids (such as acetic, propionic, and butyric acid) and bacterio-cins that communicate with the brain and other vital organs via the bloodstream. New research shows that our gut bacteria also affect our skin in this way, have an anti-inflammatory effect, and combat bad bacteria.

The gut–skin axis manifests itself in a variety of ways. Acne sufferers, for example, have more Bacteroidetes bacteria in their bowels than those with normal skin. People who are prone to eczema have fewer bifidobacteria than the norm, as well as lower levels of *F. prausnitzii* bacteria.

Professor Lars Engstrand investigates the microbiome. He and his team have observed that eczema patients have lower levels of Bacteroidetes in the gut. People with rosacea are more susceptible to infections caused by *h. pylori*. Psoriasis sufferers, for their part, have bacterial DNA in their bloodstream, indicating a leaky gut. A leaky gut is caused by an imbalance

in the microbiome that has a bad effect on the mucous membranes of the bowel. Substances that should normally remain inside the gut leak out, triggering abnormal processes in the body and the skin. A link has been observed between a leaky gut and various skin disorders, such as atopic eczema and dermatitis herpetiformis (DH), another type of eczema, which manifests as reddish bumps and blisters under the skin and which is often associated with gluten intolerance and other autoimmune disorders.

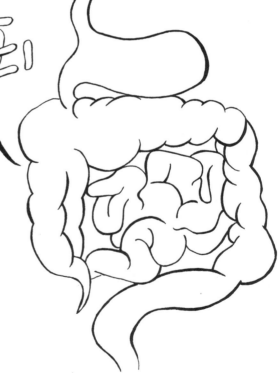

Gut bacteria produce metabolites such as short fatty acids, which in turn affect the skin. This is one aspect of the gut–skin axis.

At Birth

In their recent research, Ina and Lars have investigated what happens if babies born by C-section are swabbed with the microbiome of the mother's vaginal secretions (vaginal seeding). It's hypothesised that this is a way to prevent allergies or eczema from developing, as we pick up most of our microbiome when we're being squeezed out of the birth canal. Babies are colonised by micro-organisms from the mother, partly from her vaginal secretions, but also from her anus. After birth, the baby's skin is colonised further through breastfeeding. At the same time, micro-organisms from the baby's mother try to colonise the baby's scalp and specific areas around the child's genitals and mouth. Some of these organisms manage to achieve a healthy relationship with the human host cells, resulting in the successful transfer of a microbiome from one generation to the next.

The gut microbiome is thought to settle down by the age of three. In contrast, the skin microbiome undergoes a radical change during puberty, often accompanied by clearly visible signs in the skin. The microbial tug-of-war in the skin subsequently continues throughout the individual's life, with a shifting balance between different kinds of bacteria.

> **SAY HELLO TO LARS AND INA**
>
> Two people who know a great deal about the important role the microbiome plays in our health are Professor Lars Engstrand and Dr Ina Schuppe Koistinen from the Karolinska Institute. Lars is one of the world's leading scientists and doctors, specialising in the microbiome of the gut and other areas. Ina is a specialist in the vaginal microbiome and has a long career in the pharmaceutical industry behind her. Together they have set up the Centre for Translational Microbiome Research at the Karolinska Institute. Ina, who is also an artist, painted the attractive watercolours in this book.

From Your Toes to Your Nose — The Variable Microbiome

The flora of the microbiome varies considerably from one part of the body to another. Certain types of bacteria live on the face, while other species thrive in the navel or on the feet.

Depending on the type of bacteria found in different parts of the body, those parts can be divided up into three zones — moist, greasy, and dry. The moist areas include the navel, the groin, the soles of the feet, the inner side of the knee, and the crook of the elbow. These are home to bacteria such as *Staphylococcus* and coryneform bacteria.

There is less variety in greasy areas such as the forehead, the outsides of the nostrils, and the area behind the ears. The families that dominate here are the propionic bacteria, known as *anaeroba* because they thrive in oxygen-poor environments. So it's no wonder that they feel at home in the sebaceous

glands, which have low levels of oxygen.

The dry areas of skin, such as the skin under the arms, down the side of the body, and on the buttocks, have been shown to host the greatest variety of bacteria, with the main four families being represented to differing degrees. The face microbiome has been least researched, contrary to what you might expect, given that skincare focuses so much on the face.

We'll come back later to the connections between these bacteria and various skin problems, and to the role that greasy skin plays in conditions such as acne.

Town Versus Country

The composition of the microbiome depends essentially on a combination of heredity, diet, lifestyle, and the geographical area we live in. So we know that the skin microbiome differs from one part of the world to another.

Western people's contact with micro-organisms is extremely limited today compared with the natural interaction between humans, animals, and plants that was the norm just a century ago. In contrast to earlier times, when we lived together in closer contact with a group of individuals, today, in urbanised areas, we increasingly live in single households, several floors up, at some distance from animals and plants. We spend most of our time inside, in contact with sterile surfaces. We take exercise, but often indoors. We commute to work by tube, bus, or car.

Comparative studies of the skin microbiomes of American and Tanzanian women have shown that the former are home to more propionic bacteria, staphylococci, and streptococci, while the latter host more soil bacteria, such as rhodobacteraceae and nocardioidaceae. We know that Americans generally spend more time indoors, in contact with dry surfaces, while Tanzanians generally engage in more activities out of doors, in contact with soil and water. What about you? Is your lifestyle closer to that of an American or a Tanzanian woman?

The bacterial flora of soil, land, and animals differs from our own, and contact with it heightens our own resistance and boosts our immune system. Research shows that the growing incidence of autoimmune diseases is associated with an increasingly urban way of life. In contrast, fewer children in dog-owning families suffer from eczema. Moreover, the microbiome of dog-owners is strikingly similar to that of their pets.

Permanent Residents or Temporary Guests

Commensal bacteria are those that live permanently on or within us. Collectively, they're regarded as good bacteria, protecting us against attacks by pathogenic bacteria (the baddies). The group also includes bacteria that can be either good or bad, depending on the balance within the microbiome. Two such are *S. epidermidis* and *P. acnes*.

Just as we've been trying our best to eradicate bacteria over the last 100 years, we've also been doing our utmost to identify negative links with most of them. Yet bacteria are much like human beings in some ways: if we're torn out of our familiar surroundings, if our nearest and dearest are taken from us, if we're traumatised, we can change, and our defining features can be turned against us.

If there's an imbalance in the microbiome, commensal bacteria that are normally good for us can change sides and become opportunistic, meaning that they start to multiply and act as pathogens. This state of affairs provides favourable conditions for skin disorders such as acne, psoriasis, eczema, and rosacea.

Microbes from other animals, plants, or the soil are not part of our normal microbiome. They're transferred to us when we dig over a plot of land, cosy up to our pets, or spend time out of doors. When these microbes settle on our skin they can affect our commensal bacteria, either as uninvited guests who cause difficulties, or — more often — as friendly guests who help keep the long-term residents in good spirits.

BACTERIA — FRIENDS OR FOES?

We have a deep-rooted tendency to view these microscopic organisms as evil. This is an attitude we've inherited from the nineteenth century: for a long time, our culture has regarded bacteria mainly as enemies to be eradicated. Popular online forums — often the most accessible sources of information for the majority — show that bacteria still have an unenviable reputation.

THE BACTERIAL REVOLUTION

Bacteria and Fungi in the Skin

The skin's micro-organisms protect the skin against attacks by bad bacteria and viruses. A disruption in the balance between the commensal species can lead to disorders such as acne and eczema. The most common families of bacteria living on the skin are Actinobacteria, Firmicutes, Bacteroidetes, and Proteobacteria. Among the different types of fungi present, *Malassezia* species predominate.

Actinobacteria

The Actinobacteria family includes propionibacteria. One well-known type of propionibacteria, mentioned more often than others in the skin context, is *P. acnes*, which was long believed to cause acne. All of us have *P. acnes* in our skin, where it is part of our normal bacterial flora; its functions include the important task of protecting our skin against harmful bacteria. *P. acnes* is the predominant type of bacteria in greasy skin. Research into the T-zone (the forehead, nose, and chin) has shown a link between *P. acnes* and improved skin hydration. A Swedish research team in Lund, headed by Rolf Lood, recently showed that *P. acnes* also produces a unique enzyme that protects the skin against oxidative stress as well. This enzyme has just as powerful an effect as vitamins C and E.

If the bacterial balance of the microbiome is out of kilter, however, *P. acnes* can be a contributory factor in inflammation and acne. Acne can break out when *P. acnes* produces free fatty acids in the sebaceous glands. As a result, by-products form that risk irritating the sebaceous follicle and setting off an inflammation that ultimately erupts as an inflamed skin rash.

Corynebacteria also belong to the Actinobacteria family and are considered to be part of our normal skin flora. One example is *C. jeikeium*, which thrives not only in the skin, but also in soil and water. It has been suggested that it is an important antioxidant, as it absorbs manganese in the skin, which helps to counteract oxidative stress.

Corynebacteria are used in the cheese-making process. If you enjoy mature cheese, you might like to know that corynebacteria help form its rind. Once you realise that, it may seem less odd that these bacteria are also responsible for the sometimes less-than-pleasant aroma exuded by armpits!

Firmicutes

Out of the family of Firmicutes bacteria, staphylocci are those most relevant to our skin. *S. epidermidis* is the most common type of bacteria in our skin, where it is probably accounts for over 90 per cent of the total bacterial flora. An interesting study conducted by research teams in various countries, including the USA and Germany, has shown that *S. epidermidis* living in and on our noses protects us against the flu virus.

So does that mean that cutting down on personal hygiene and washing less — not keeping our noses clean, so to speak — will help us avoid the flu? We can't be sure, but it's an intriguing theory.

S. epidermidis is a very good friend to most of us, helping to preserve the bacterial balance that's important in maintaining a healthy skin by working together actively with the skin's immune defences to keep pathogenic bacteria at bay. It also supports and enhances skin hydration.

Moreover, *S. epidermidis* can protect against acne by striking a balance between *P. acnes* and *S. aureus*, a streptococcus. However, glycerine is needed for this to happen. Glycerine occurs naturally in the skin, but as you may recall from Chapter 2, it's also one of the most important humectants in skincare products. So there are two good reasons to use it.

A new study has put forward the hypothesis that *S. epidermidis* may also protect against skin cancer.

S. AUREUS

Like *S. epidermidis*, *S. aureus* belongs to the staphylococci group and is found mainly in the nasal cavity, where it belongs to the nose's commensal bacteria. *S. aureus* is present in at least 30 per cent of healthy people. It isn't harmful, but lives in symbiosis with human host cells — in most cases. Under certain circumstances, however, *S. aureus* can become a pathogen and cause infections. It plays a pathogenic role in acne, rosacea, psoriasis, and eczema. Infections caused by *S. aureus* are generally treated with antibiotics, but more and more resistant varieties are developing, which don't respond to such treatment. As an alternative to antibiotics, scientists are looking at how *S. epidermidis* can be used to displace *S.aureus* and at ways of restoring balance to the microbiome with the help of phages, which we will examine in more detail later.

In a study conducted by a big cosmetics firm in cooperation with a hospital in the French city of Nantes, microbiome samples were collected from twenty-six individuals with acne. Proteobacteria and Firmicutes were found to predominate in their bacterial flora, which also contained a smaller proportion of Actinobacteria, including *P. acnes*. Their spots, on the other hand, were found to contain far more of the staphylococcus bacterium *S. aureus* than the surrounding healthy skin. *P. acnes* actually accounted for less than 2 per cent of the entire microbiome.

Bacteroidetes and Proteobacteria

Unlike the gut, where Bacteroidetes predominates — accounting for approximately 90 per cent of the microbiome — the skin has far less of the bacterium, though it does occur. There has been relatively little research so far into Bacteroidetes' role in the skin microbiome, but new findings are emerging all the time, and French scientists have recently identified a type of Bacteroidetes that grows only on the skin. It has been given the name *B. cutis*, cutis being the word for skin in medical Latin. Bacteroidetes seems to have a particular significance for the skin. Lars Engstrand and his team have found that eczema-sufferers have less Bacteroidetes in the gut. Bacteroidetes has also been shown to have an anti-inflammatory effect on the skin.

Proteobacteria include the group Acinetobacter, which studies have shown to provide protection against allergens with which we may come into contact via the skin.

Each type of bacteria in this group is thought to play a specialised role in the skin: anti-inflammatory, anti-oxidative, humectant, or anti-allergenic. As the skin microbiome is mapped in increasing detail, more bacteria are being renamed in a way that reflects their function in the skin. Examples include *B. cutis*, mentioned above, and *P. acnes*, which is increasingly referred to as *Cutibacterium*, or, in the short form, *C. acnes*.

Can You Analyse Your Own Microbiome?

Yes — plenty of firms will analyse your gut microbiome for you. Based on the composition of your bacterial flora, they'll advise you on what to eat. There are even some firms that analyse the skin microbiome these days — but it's not yet feasible to make recommendations about diet and skincare on the basis of their results.

One day we're sure to see clinics employing doctors (gastroenterologists and dermatologists) and dieticians to diagnose the gut and skin microbiomes. Then they'll put the results together to provide an overall assessment and recommendations on diet, skincare, and lifestyle. Bacteria run our lives — we just need to get used to it!

Yeasts

The *Malassezia* species yeast dominates the skin mycobiome — the fungal answer to the bacterial microbiome — and can be found in nearly all healthy people. A dramatic increase in the incidence of this yeast has been observed in skin conditions such as eczema, seborrhoeic eczema, acne, and other inflammatory disorders. Yeasts are thought to secrete substances that, in their turn, trigger the immune system, resulting in inflammation.

As we've seen, past studies have tended to focus on *P. acnes* as a cause of acne. In recent years, it has been realised that there's a different kind of acne, caused by the yeast *Malassezia*. Very recently, an interesting study carried out in Turkey has shown that, of over 200 patients who sought medical care for their acne, over a quarter had malassezia folliculitis, an inflammation of the follicles that resembles acne.

If there's an imbalance in the skin microbiome, yeasts seize the opportunity to spread. This condition may take the form of whiteheads on the face, throat, chest, back, or the outer side of the arms.

Malassezia folliculitis must not be treated with antibiotics, which can actually make matters worse. The right way to tackle it is to use fungicides — just as for athlete's foot.

The same applies to eczema, which can also be caused by yeasts. This specific disorder is known as *seborrhoeic eczema*.

Eczema and Seborrhoeic Eczema

Eczema is the best-known case of a link between an imbalance in the microbiome and a skin disorder. This condition has become increasingly common, doubling or even tripling over the last few decades in industrialised countries. This may be linked with our way of life: keeping things too clean, taking too many antibiotics, and reducing our contact with animals and the natural world. In eczema sufferers, an increase in *S. aureus* in the skin has also been observed, with a corresponding reduction in propionibacteria and coryne-bacteria.

Higher levels of yeasts have also been observed in the area affected by eczema. Seborrhoeic eczema is a condition which many associate with the scalp, but this type of eczema also appears on the face and other parts of the body.

BACTERIA AND FUNGI IN THE SKIN

Johanna Introduces...

Lars Engstrand
Gut Microbiome Specialist at the Centre for Translational Microbiome Research, Sweden

I've had the pleasure of meeting microbiome specialist Lars Engstrand many times in the last couple of years. Though most of our discussions have been about the skin microbiome, Lars's specialist subject is actually the microbiome of the gut. He was a member of the scientific team that examined the stomach contents of Ötzi the Ice Man, who lived over 5000 years ago.

How do you view the link between the gut and the skin microbiome?

There's been a great deal of progress over the last few years as regards the skin microbiome. Research in this area has some way to go to catch up with research into the gut microbiome, but it's clear there's a lot of interest in the skin. Just as with chronic inflammation of the bowel, there seem to be fewer bacteria that secrete short fatty acids when someone has a skin inflammation.

How might we use bacteria to treat skin disorders?

First we need to establish which bacteria are best for a healthy skin. Then we might consider cultivating these bacteria and applying them to skin whose bacterial balance has been disrupted. This is exactly what's already being done with the gut microbiome — partly through faecal transplants (transfers of faeces) and partly by transferring bacteria in tablet form.

I'm sure we're going to see similar methods being used to benefit the skin.

What is it about micro-organisms that fascinates you?

I've been working in various areas of microbiology for thirty-odd years now. During the second half of my career, I've got more and more interested in inflammatory bowel diseases. We realised a long time ago that the bacterial flora of patients with these illnesses was out of kilter. That prompted us to improve our understanding of how bacteria affect our health, and how they affect each other.

Together with my friend and colleague Martin J. Blaser, I've researched *Helicobacter pylori* bacteria, which are associated with stomach ulcers. Half the people on this planet have these bacteria in their stomachs, but we obviously don't have half the population suffering from stomach ulcers. In fact, we've started to see *Helicobacter*

Cultivated skin bacteria are the future.

pylori as an asset — it's actually been shown to provide protection against more dangerous bacteria, such as the ones that cause cholera. It's all a question of balance.

Is there anything you can point to as the next big thing?
A year or so ago I attended an interesting conference in Paris, at which some scientists described how they'd been trying to solve the puzzle of why horses roll about on the ground. They studied soil samples and discovered a bacterium (*Nitrosomonas eutropha*) that colonises horses and lives on their sweat — or on the ammonia in their sweat, to be precise.

Johanna: Yes! A skincare firm has tried adding these bacteria to water and spraying people with the liquid. The people who've tried this say they don't get smelly, even after they've stopped washing with soap. Now scientists are wondering whether modern

societies have become so obsessed by hygiene that we've got rid of *Nitrosomonas eutropha* — a bacterium that was good for us and played an important role in our skin.

Thank you, Lars!

Boost Your Microbiome

There are two ways to boost the good bacteria in your skin — by exploiting the gut–skin axis, or by transferring bacteria directly to your skin. You can support the balance in the skin microbiome with the help of fibre, live bacteria, and substances secreted by bacteria. Today's skincare manufacturers have recently become aware of how important bacteria are and are cautiously beginning to make use of them in their products.

BOOST YOUR MICROBIOME

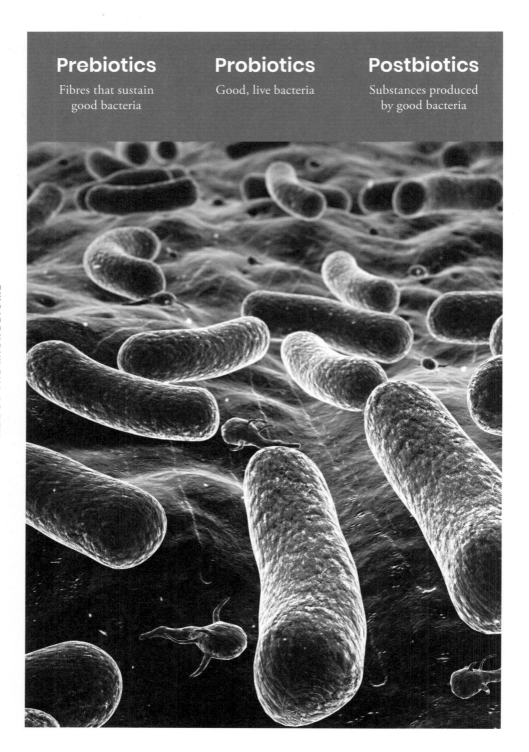

Prebiotics
Fibres that sustain
good bacteria

Probiotics
Good, live bacteria

Postbiotics
Substances produced
by good bacteria

ABOUT THE MICROBIOME

Biotics

The term 'biotics' covers prebiotics, probiotics, and postbiotics, all of which are used to enrich our microbiome. Our skin is affected both by what we put on it and, through our gut, by the food we eat. As regards the impact of biotics on the skin via the gut, probiotics have played a dominant role, while conventional skincare has so far focused on postbiotics. But times are changing, and it's likely that skincare will also be dominated by probiotics soon.

Prebiotics for the Gut

Prebiotics, which are found in fibre, feed our bacterial flora. When fibre reaches the large intestine, it boosts good bacteria, enabling them to grow and multiply. There are particular types of fibre that provide especially good prebiotics. Look out for inulin, glucomannan, oligosaccharides, and beta-glucan. These substances can also be found in our food, especially in onions, garlic, Jerusalem artichokes, unripe bananas, maize (corn), rolled oats, and cold potatoes. A recent Korean study showed that people who ate one gram a day of the prebiotic galacto-oligosaccharide for three months had better-hydrated skin and fewer wrinkles. Galacto-oligosaccharide can be found in foods including beans and seaweed. Polyphenols, a group of anti-oxidant substances, also have a prebiotic effect. We'll look at this in more detail in the chapter on lifestyle.

Probiotics

Probiotics means actual live micro-organisms. You can find probiotics in everything that's fermented: filmjölk (a curdled milk product popular in Scandinavia), yogurt, pickled vegetables, kombucha, tempeh, and miso (read more about this in the chapter on food). It's also possible to cultivate various kinds of lactobacilli and prepare them in concentrated form. If you take dietary supplements to get your probiotics, what you're eating is lactobacilli.

TREATING SKIN DISORDERS

In vitro studies of skin cells have shown that probiotics can relieve eczema, acne, and rosacea. Most such studies have involved using various types of probiotics in an attempt to limit the growth of *S. aureus*. The species that have proven most effective against eczema are: *L. reuteri, L. rhamnosus GG, P. freudenreichii, P. acnes, L. paracasei, L. acidophilus, L. casei, L. plantarum, L. bulgaricus, L. fermentum* and *L. lactis*.

BOOST YOUR MICROBIOME

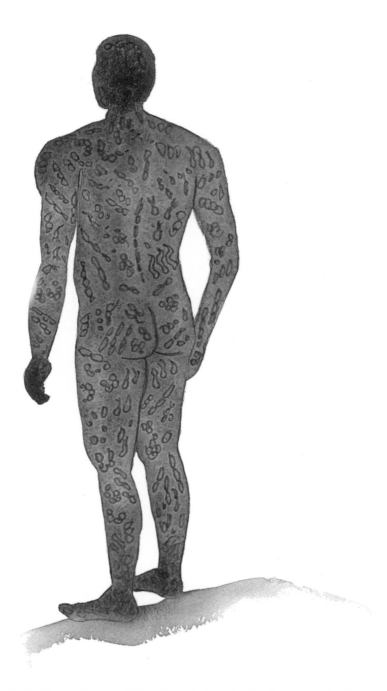

The bacterial flora that inhabit our skin are invisible to the naked eye. But they're very much alive. In fact, our skin itself is alive with micro-organisms — most of which are good.

ECZEMA

Clinical studies on human subjects have also demonstrated that probiotics can alleviate eczema. For example, a Korean study of children with eczema showed that half the children responded to treatment within three months of taking *L. plantarum*. The children who responded positively had relatively active eczema involving a large number of inflammatory factors. However, there are also studies of probiotics that show no effect on eczema. The simple fact is that more studies are needed to establish whether bacteria really do affect eczema, and, if so, which species are effective.

SEBORRHOEIC ECZEMA

Fungicides have long been the most common antidote to yeast infections. However, new studies show that consuming probiotics can also help displace yeasts and improve the condition of patients with seborrhoeic eczema. Probiotics with *L. paracasei* bacteria have proven effective here.

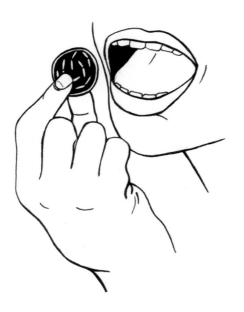

ACNE

The first clinical studies evaluating the effects of probiotics on acne were carried out by Robert H. Siver, a doctor, in 1961, and involved various families of lactobacilli. He suggested that there were interactions between acne and processes in the intestinal canal. This observation confirms what we today call the gut–skin axis.

Most of the studies carried out so far have been done in vitro. However, they've at least shown that it's possible to inhibit the growth of *S. aureus* and *P. acnes* bacteria with the help of *Lactococcus sp. HY 499* and *L. paracasei*.

An ongoing clinical study of *Nitrosomonas eutropha* soil bacteria is investigating the probiotic treatment of mild to moderate acne. The treatment reportedly involves utilising the bacterium's nitrogen cycle to convert ammonia and urea, found naturally in human skin, into nitrite and nitrogen oxide. These substances possess anti-inflammatory and anti-microbial properties.

At the time of writing, only one study has shown probiotics to have a positive effect on acne. In 2016, a team of scientists from Italy showed that women acne sufferers who took *L. rhamnosus* experienced an improvement in their condition.

ROSACEA

Rosacea has also been shown to respond to some extent to treatment with probiotics. In 2014, the American Academy of Dermatology suggested using probiotics to treat the condition. This is because probiotics affect the immune system and relieve skin inflammation. However, there have been fewer studies of rosacea than of acne and eczema.

BOOST YOUR MICROBIOME

ETERNAL YOUTH?

The notion that probiotics offer some magic formula for eternal health and beauty needs to be taken with a pinch of salt. However, there are studies that have shown a reduction in oxidative stress and inflammation in the skin, plus an enhanced barrier function and a better-hydrated skin, as a result of consuming probiotics.

In another study, which looked at the gut–skin axis in Japanese women students, participants drank probiotic-curdled milk containing *L. breve* bacteria for four weeks. The results showed more moisture in the skin, an improved uptake of nutrients, an increase in the volume of faeces excreted, and a reduction in wrinkles. At the same time, the secretion of sebum in the skin increased.

Scientists from South Korea have also shown in a study that taking *L. plantarum* can improve skin hydration, give the skin more of a glow, reduce wrinkles, and firm the complexion. Although we're only talking about a small number of studies here, these are interesting observations.

In addition, consumption of the probiotic lactic acid bacterium *L. paracasei* has been shown to reduce skin sensitivity. The study concerned involved triggering the TRPV1 sensitivity receptor with capsaicin from chili peppers. Participants experienced a significant reduction in skin sensitivity when their skin was exposed to the substance.

Biotics in Skincare

Microbiotic skincare is an area in which product developers have just begun to take an interest. Have you noticed that quite a few skincare products are now being marketed as probiotic? Unfortunately, they're often marketed incorrectly. Most products that claim to contain probiotics actually have prebiotics or postbiotics in them. Prebiotics — inulin, glucomannan, and oligosaccharides — can be found in several skincare products. Their purpose is to keep the skin balanced and enrich the skin's microbiome. Application of prebiotics to the skin has been shown to improve skin status.

INCI: Inulin, Glucomannan, Alpha-Glucan Oligosaccharide

Postbiotics are bacterial extracts containing substances produced by probiotics. Although they're not the same as probiotics, certain bacterial extracts have been shown to have a positive effect on the skin. For instance, a bacterial extract of *V. filiformis*, a marine bacterium, can alleviate eczema. Similarly, if postbiotics derived from *L. plantarum* are applied to the skin, it seems to have a beneficial effect on acne.

To identify postbiotics, look out for 'ferment' and 'lysate'. Examples include:

INCI: Lactobacillus Ferment, Lactobacillus Lysate, Bifidobacterium Lysate.

Genuine probiotic skincare has recently emerged on the market, and a few manufacturers put live bacteria in their creams. The probiotics they use are mostly lactobacilli and *Nitrosomonas eutropha*. Using live bacteria poses significant challenges for manufacturers.

Come Down to Earth Again

We know that a wide variety of bacteria is good for the skin. One way to increase their diversity is to copy what horses do — roll around on the ground!

In a recent Finnish study, healthy volunteers touched sixteen types of soil, plant-based material, and moss. Samples were taken from the participants' microbiome before and after their brief contact with these substances. The study showed an increase in the variety of organisms in their skin microbiome.

The reason for this rather odd study was concern about the rising incidence of autoimmune diseases over the last few decades. The hygiene hypothesis suggests this is linked with higher levels of hygiene and reduced contact with natural biodiversity.

Will we reach the point where we rub soil into our skin, cover ourselves in moss, or go on forest 'spa retreats' that involve rolling on the ground? One skincare company has gone in this direction by selling sprays containing *Nitrosomonas eutropha*, a type of bacteria found in the soil. The way this is believed to work is by bacteria breaking down ammonia (from our sweat) and converting it into nitrogen, which has both anti-inflammatory and antimicrobial properties. Clinical testing is ongoing on how these soil bacteria affect skin disorders such as acne, eczema, and rosacea — plus migraine and high blood pressure, another two disorders typical of Western societies.

MAJOR CHALLENGES FACING MANUFACTURERS

Using probiotics in skincare is tricky and poses major challenges. There are two main reasons that probiotic products on the market are so few and far between.

1. Skincare products contain water, and it's hard for bacteria to survive in water-based products.

2. These products' shelf life is a major issue. Skincare products have an established shelf life of up to thirty months. But there's no way to guarantee that any bacteria in a product will still be alive after such a long time, particularly if the product is stored at room temperature. Compare this with probiotics in food products; it's always recommended to keep them at low temperatures. On top of that, it's hard to use preservatives in probiotic skincare, as the whole point of them is to kill bacteria.

Two possible solutions to these problems are to make these skincare products completely water-free or, alternatively, to put the probiotic content in capsules. These are the two methods currently in use.

BACTERIA FOUND NATURALLY IN THE SKIN

Before very long we'll be seeing probiotic skincare products based on bacteria that live naturally in the skin. These will be based on the idea of transferring skin bacteria from individuals with healthy skin to patients with a skin disorder.

The first study into the technique of transferring skin bacteria is already a fait accompli. Eczema sufferers were sprayed with skin bacteria (*Roseomonas mucosa*) from people with normal skin. The study showed a noticeable improvement in their condition within a few months, provided that they were sprayed twice a week. This interesting study was carried out in the US in 2018. But it's important to stress that it was only done on a small scale. More and larger studies are needed to confirm the results.

And what would you say to a day cream containing *S. epidermidis*, which inhibits the growth of the bacteria responsible for acne and eczema? An American start-up is developing probiotics based on *S. epidermidis* that are designed to be sprayed on or rubbed into the skin. At the time of writing, the firm is conducting clinical tests, and we don't yet know the outcome. But this is the same technique that's been used for some time in transplanting faeces to the gut so as to restore balance to the microbiome. It has achieved excellent results in treating a diarrhoeal illness caused by *c. difficile* bacteria.

Skincare — Present and Future

Research into the skin microbiome is still in its infancy. As we've seen, the gut microbiome affects the skin, but the skin's own microbiome, too, is more important to skin health that we could ever have imagined.

Instead of deploying anti-microbial agents, preservatives, or antiseptic cleansing agents to eliminate our microbes at all costs, we ought to regard our bacteria as friends and treat them accordingly. Maybe we should be adding probiotics to expand our circle of friends, or inviting the commensal bacteria that are good for our skin for a tasty prebiotic meal? Just as it's a bit off to start cleaning and hoovering the dining room while your guests are still at table, it's a bit off to bathe our friendly skin bacteria in all the antimicrobial substances we apply daily to our skin.

Maybe changing attitudes will pave the way for preservative-free skincare based on bacteria that occur naturally in the skin. It looks as if a whole new outlook on bacteria and skincare is on its way!

PRESERVATIVES

We still don't know enough about how preservatives affect the gut microbiome. In a study conducted by the University of Hawaii, preservatives were examined in relation to disrupted gut flora, with a focus on two common food peeservatives: sodium bisulphite and sodium sulphite. These were tested on four known species of 'good' bacteria that are commonly found in the natural gut flora: *L.casei*, *L. plantarum*, *L. rhamnosus*, and *S. thermophilus*. What happened? The preservatives killed the good bacteria.

Although the significance of this result isn't yet completely clear, such studies may well change the dominant view of preservatives as beneficial, non-toxic antimicrobial compounds.

Large quantities of preservatives (parabens, phenoxyethanols, glycols, and alcohol) are added to skincare products to extend their shelf life. You don't have to be a skin scientist to understand that preservatives harm the skin microbiome. Why should an antibacterial substance like a preservative kill only bad bacteria? At the time of writing, Chinese scientists have demonstrated exactly that. All of the five preservatives most often used in skincare reduce the levels of good bacteria in the skin. It's interesting to speculate whether the recent increase in skin complaints like acne is a result of using preservatives on a large scale in skincare products.

At the Skin Microbiome Congress held in Boston in 2018, one of the central issues was the extent to which the synthetic chemicals we apply to our skin disrupt the microbiome, and how skincare should help improve the skin's balance instead.

The same issue was on the agenda of the 2019 Skin Microbiome Congress. The material made available stated that today's average skincare consumer in the US spends about $1.50 a day on skincare, exposing themselves to as many as 500 chemicals in the process. It's predicted that the revolutionary skincare of the future will take proper account of the gut and skin microbiome.

TREATING ACNE WITH FAECAL BACTERIA

One of the most fascinating studies of a skin disorder for a long time was carried by a South Korean research team. A skin lotion containing added faecal bacteria extracted from the stool of healthy people was tested on seventy patients over the age of twelve with mild to moderate acne. Participants were divided into two groups. Half the people were told to apply the cream twice daily, while the other half were given a placebo cream without any probiotic content. The team found less skin inflammation in the first group. The probiotics were also seen to help achieve a balance with *p. acnes*. Cell studies have shown that probiotics (*s. salivarius* and *e. faecalis*) inhibit the unhealthy growth of *p. acnes*. One day these bacteria may be used as an ingredient of a probiotic cream to treat acne.

BOOST YOUR MICROBIOME

Phage Therapy Instead of Antibiotics

Another likely future scenario for our skin microbiome is the rise of phage therapy. This involves deploying bacteriophages, viruses that can kill many of our most harmful pathogenic bacteria (mostly very efficiently and in a carefully targeted way). Most phages kill their host bacterium quickly and efficiently, within an hour or so, but before the host dies, it has produced large quantities of new phage particles which can then infect other bacteria.

It's possible to develop phages that target one specific species of bacteria in the skin, such as *S. aureus* or *P. acnes*. Instead of tackling acne with antibiotics, we could avoid both antibiotic resistance and the side-effects of antibiotics (which kill off good bacterial flora) by treating acne with phages. However, as we've seen, research in this field is still in its infancy.

Thaw Out Your Own Youthful Bacteria

The composition of our skin's bacterial microbiome seems to change as we age. The proportion of actinobacteria declines, while the proportion of proteobacteria increases. An American cosmetics firm has investigated which bacteria increase proportionately as we age, and they think it might be possible to treat ageing skin with a special youthful strain of corynebacteria so as to reduce wrinkles and make the complexion look more youthful. It's not yet clear quite how they plan to do this, but I think we'll soon be seeing plenty of useful new products for mature skin on the market.

Once, when Professor Lars Engstrand and I were speculating about the future, we agreed there'd soon be clinics for young people to freeze their healthy young skin microbiome. When they're older, the microbiome can be thawed out and the bacteria cultivated. Then, all they'll have to do is apply their own thirty-year-old bacteria to their skin.

ABOUT THE MICROBIOME

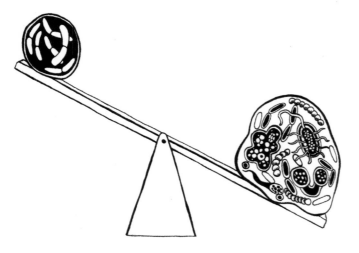

About Lifestyle

Can you eat, run, or sleep your way to a healthier skin?

Everything you put in your body, from carrots to tobacco smoke, affects your skin. You can make your skin healthier by exercising — or unhealthier by exposing yourself to stress. Where you live matters, too. If you live close to nature (surprise, surprise!) you'll have healthier skin than if you live in a mega-city with massive air pollution. In fact, our lifestyle is thought to account for 75 per cent of our skin status, while our genes account for a mere 25 per cent.

Food for a Healthier Skin

As a skin scientist, I'm often asked if the food we eat can affect our skin.

The answer is a resounding yes.

Though I'm fond of cooking and eating, I'm not an expert in the specific nutrients found in different foods, or in how much we need daily of each kind. So I asked dietician Karin Magnusson to help me with this chapter. Together we've developed a nutritional guide that's designed to maximise benefits to the skin. Now we'll take a closer look at how each nutrient affects our skin.

NUTRIENTS

All nutrients are transported by the bloodstream from the gut to the skin. Nutrients well known to be good for the skin are vitamins A, C, D, E, various forms of vitamin B, zinc, selenium, polyphenols, carotenoids, omega 3 and omega 6, plus amino acids.

These days more and more people are eating less varied diets. Eat as much as one potato at lunch, and you may well feel obliged to tell your workmates you've booked a cross-fit session to compensate. As a result, we're at greater risk of nutrient deficiency than in many a long year. This deficiency can affect our bodies in various ways. We may feel more tired, look paler, and have less healthy nails and hair. And it goes without saying that our skin suffers.

Eat a varied diet and you'll automatically get all the nutrients you need — apart from vitamin D, which is a bit trickier.

But if you exclude one or more food groups, you'll need to do your homework properly. If you choose not to eat meat, fish, and dairy products, for example, you'll need to pay particular attention to your intake of zinc, iron, selenium, calcium, and vitamin

B12. Cutting down on empty carbs like soft drinks, white bread and confectionery is great, but if you stop eating fruit, pulses, root vegetables, and wholegrain cereals, you may end up with too little fibre in your diet. It's also worth bearing in mind that many types of fibre are prebiotic, making them an excellent source of food for the good bacteria in your gut. And as you'll recall, a healthy gut goes with a healthy skin.

If you're in good health and eat a variety of foods, there's no need to worry about dietary deficiencies. However, some people may find themselves in the risk zone: those suffering from chronic gastrointestinal disease or untreated gluten intolerance, for example. And if your diet excludes certain food groups, you may need to review your eating habits.

Vitamin A and Beta-Carotene — Anti-Ageing Molecules in Your Food

The right dose of vitamin A is important for the skin in several ways: it stimulates collagen formation in the dermis and supports the renewal of the epidermis. The amount of collagen in the skin tends to decline as we age, while epidermal regeneration slows down. This means that foods rich in vitamin A and beta-carotene are a kind of antidote to ageing. Vitamin A is found mainly in foods of animal origin, but there's also a handy alternative — beta-carotene.

Beta-carotene, which belongs to the group of antioxidants known as carotenoids, is converted to vitamin A in the body when needed. When you take in beta-carotene from carrots (the very best source), your liver takes two to four hours to produce vitamin A, which is then transported to your skin.

The other way in which beta-carotene is relevant to skin is its effect on acne. In one study, scientists compared the amount of vitamin A and E in the blood of 100 acne sufferers and 100 control subjects without acne. The acne sufferers proved to have significantly lower levels of these vitamins than the control subjects. Acne patients are often treated medically with vitamin A acid, whose active substance, isotretinoin, prevents the sebaceous glands from producing too much sebum.

Vitamin A is fat-soluble and is stored in the body. That means there's a risk that levels may rise too high. Excessive doses have been shown to have effects ranging from liver damage and light sensitivity to hair loss.

This is how much vitamin A you need daily:

Women: 700 RE
Men: 900 RE

You'll find retinol equivalents in these foods:

One egg: 125 RE
Two slices of cheese: 82 RE
125g of liver: 6700 RE
One carrot: 600 RE
70g of curly kale: 185 RE
125g of Hippophae rhamnoides (sea buckthorn): 133 RE
100g of papaya: 87 RE
100g of pumpkin: 87 RE

FOOD FOR A HEALTHIER SKIN

WHAT IS RE?

RE stands for 'retinol equivalents'. Retinol equivalents are units of active vitamin A from both vitamin A-rich foods and from beta-carotene (which, as we've seen, is converted into vitamin A in the body).

More Carotenoids That Are Good for the Skin

Other carotenoids besides beta-carotene also protect the skin. About fifty different kinds can be found in our food. They include alpha-carotene, zeaxanthin, lutein, astaxanthin, and lycopene. All of these reduce the levels of free radicals and oxidative stress in the skin.

Astaxanthin is a carotenoid with strong antioxidant and anti-inflammatory properties. It's found in lobster, salmon, microalgae — and flamingos. (Astaxanthin is what makes them pink.) Nearly all of the research on this antioxidant has been carried out in Japan. A study with sixty-five female participants was carried out in 2017. For four months, they were given either a dose of up to 12mg of astaxanthin or a placebo. Those who took astaxanthin had fewer wrinkles, and their skin was better hydrated. In a study carried out the following year, the dosage was reduced: twenty-three healthy Japanese participants took either a capsule containing 4mg of astaxanthin or a placebo over a period of ten weeks. The participants who took astaxanthin were seen to become less sensitive to light, and their skin was better hydrated. They also rated themselves as having smoother skin. The researchers concluded that astaxanthin really does help improve skin health.

In a German study, women who took daily carotenoid supplements (lutein, beta-carotene, lycopene, and zeaxanthin) for ten months increased the concentration of carotenoid in their skin, resulting in a stronger antioxidant effect. There was also an increase in the amount of collagen and elastin in the dermis.

Lycopenes are fat-soluble, so uptake is easier if you consume them with oil. In a recent German study, women and men who ate tomato paste (containing lycopenes) for three months were less susceptible to sun-related irritation than participants who ate no tomato paste.

You'll find carotenoids in:

carrots, tomatoes, oranges, curly kale, spinach, salmon, shrimps, and microalgae.

CAROTENOID–RICH GAZPACHO FOR THE SKIN

One finely grated carrot
Four cherry tomatoes
One tablespoonful of cold-pressed olive oil
One clove of garlic
200ml cold water
Salt and pepper

Instructions

Put all the ingredients in a mixer and switch it on full. Pour the resulting liquid into a glass. Bon appétit!

NIACIN AND THE SKIN'S BARRIER FUNCTION

Niacin, vitamin B3, is the best-studied B vitamin relevant to the skin, as it strengthens the epidermis and is therefore good for hydration.

The history of niacin began in the eighteenth century, when a serious skin complaint that's today known to be associated with niacin deficiency was discovered among Spanish farmers. This disease of malnutrition, pellagra, has sometimes been called the sickness with the four d's: dermatitis, diarrhoea, dementia, and death. Niacin deficiency is very unusual in Sweden, and likewise in the UK, the US, and Australia.

One of its symptoms is a brown, hyper-pigmented skin that is dry, hard, and rough.

Both rosacea and acne have been shown to improve when the sufferer takes niacin.

High levels of niacin from dietary supplements can result in a skin reaction called niacin flush. This reaction is characterised by a burning, hot, or stinging sensation in the face and on the chest.

This is how much niacin you need daily:

Women: 14 NE (niacin equivalents)
Men: 18 NE

You'll find niacin in these foods:

50g of peanuts: 5.6 NE
125g of roast or fried chicken: 25 NE
12g of salmon: 15 NE
One slice of wholemeal bread: 1 NE
One tablespoonful of wheat bran: 1.5 NE

VITAMIN B12 — THE BACTERIAL VITAMIN

Vitamin B12 is secreted by bacteria that we take in mainly from animal source foods. Animals eat food containing the right kind of bacteria, and B12 is transferred to us through our food. The lactobacilli in fermented products also produce B12. That's why yogurt, curdled mild (filmjölk), and fermented products such as sauerkraut are good sources of vitamin B12.

Vitamin B12 has a complex link with the skin. Various disorders, including vitiligo and mouth sores (aphthous ulcers), can be associated with a vitamin deficiency. Vitamin B12 deficiency can also result in a pale, washed-out skin. Another symptom of vitamin deficiency is hyper-pigmentation.

Older people often suffer from a vitamin B12 deficiency, as the capacity to absorb the vitamin declines with advancing age. In some cases, acne sufferers treated with vitamin A acid have been observed to be at risk of vitamin B12 deficiency.

This is how much vitamin B12 you need daily:

Adults: 2µg

You'll find vitamin B12 in these foods:

200ml milk: 1.2µg
200ml curdled milk (filmjölk) or yogurt: 0.5µg
200ml of vegetable-enriched drink: 1.2µg
125g of beef: 1µg
One egg: 1.3µg
125g of salmon: 3µg
Two slices of cheese: 0.3µg

The unit µg is pronounced 'microgram'.

FOOD FOR A HEALTHIER SKIN

FOLATE — VITAMIN B9

Folate is better known as *folic acid*. When the vitamin is found in food, it's called folate; folic acid, on the other hand, is a synthetic form of folate used to enrich foods and as a dietary supplement. If you suffer from a folate deficiency, you should take a folic acid supplement. The best way to get this vitamin is to eat raw vegetables; alternatively, keep the cooking time short or steam your vegetables.

Tiny amounts of folate are found in nearly all foods. It's needed to form new cells, especially red blood cells. The process of cell formation is faster in growing youngsters and pregnant women. That's why expectant mothers need more folate, while women who are trying to conceive have a particular need for folate-rich foods, or may require folic acid supplements.

Folate deficiency can manifest as a pale, sallow skin. A new study shows that this deficiency is most common during the summer. This is because folate is sensitive to sunlight, which makes it break down in the skin. Light-skinned people thus tend to have lower levels of folate in the bloodstream. A study carried out by Uppsala's Academic Hospital showed that a combination of vitamin B12 and folic acid can alleviate vitiligo.

This is how much folate you need daily:

Women: 400µg
Men: 300µg

You'll find folate in these foods:

110g of liver: 114µg
30g of fresh spinach: 60µg
70g of fresh curly kale: 80µg
One carrot: 16µg
80g of red lentils, cooked: 80µg
80g of chickpeas, cooked: 80µg
200ml curdled milk (filmjölk): 24µg

VITAMIN C — THE COLLAGEN BOOSTER

Vitamin C deficiency was first described in Western medicine as scurvy, a disease that emerged on the long voyages of fifteenth-century European navigators. The symptoms were bleeding gums, poor healing of sores and wounds, corkscrew-like hair growth, and thinner skin (vitamin C is essential for collagen formation in the dermis).

Once we've taken in vitamin C in our food, it's transported in the bloodstream to the dermis. Higher levels of the vitamin have been measured in the skin than in other parts of the body, indicating a clear link between vitamin C intake and skin health. This is confirmed by an American study showing that an adequate intake of vitamin C is associated with better skin quality, while a deficiency can be linked with a 10 per cent increase in the risk of developing dry, wrinkled skin. However, since there's a limit to the amount of vitamin C that can be absorbed by the blood, dietary supplements benefit only people with abnormally low levels. Smokers seem to have less capacity to bind vitamin C in the blood, but studies show that its concentration returns to normal levels if they stop smoking.

This is how much vitamin C you need daily:

Women and men: 75mg

You'll find vitamin C in these foods:

100g of blackcurrants: 75mg
10 strawberries: 60mg
One orange: 65mg
One red pepper (capsicum): 100mg
75g of broccoli: 40mg

VITAMIN D — THE SUN VITAMIN

Vitamin D has been the focus of much attention over the last decade. It's hard to make sure you have enough of this vitamin if you live somewhere with relatively few hours of sunshine. The vitamin is synthesised in the skin when sunlight reacts with a form of cholesterol. It's long been viewed as being important mainly for bone formation, but new research shows that it affects nearly all tissues, including those in our brain, heart, muscles, immune system, and skin.

A good deal of research has been done into the link between vitamin D and the skin, where it's important in regenerating the epidermis, healing wounds and ensuring the immune system works properly. Vitamin D is thought to help prevent sebum production and inflammation in skin with acne, and research has provided some support for the theory. A Turkish study shows that acne sufferers have lower levels of vitamin D in the bloodstream. An Australian research team who monitored a group of acne sufferers for ten years found that they had significantly lower levels of vitamin D in the bloodstream than the general population. However, the link between acne and vitamin D remains unclear. Recently, a research team in Iraq measured vitamin D levels in the blood of acne sufferers, but was unable to establish any connection.

If there's a link with skin disorders like rosacea, vitiligo, and eczema, it's even less well established. One study of rosacea revealed low levels of vitamin D in the bloodstream, while another showed excessively high levels. In the case of vitiligo, the results are contradictory: some results show lower levels of vitamin D among sufferers, while others show no significant difference in vitamin D levels among the subjects studied. The same results have been found among eczema sufferers, though it's been shown that giving vitamin D to individuals who had low levels of the vitamin to begin with alleviates their symptoms.

There are several groups of people at risk of vitamin D deficiency. They include those who exclude fish, eggs, and milk from their diet; children under the age of two; people who wear clothes that cover the whole body; people with relatively dark skin living in regions with little sunshine; pregnant women with insufficient vitamin D in their diet; and older people who spend little time out of doors.

This is how much vitamin D you need daily:

Adults: 10µg

You'll find vitamin D in these foods:

125g of salmon: 12µg

125g of pike-perch (walleye): 35µg

100g of marinated herring: 11µg

One egg (yolk): 0.8µg

200ml of vitamin D-fortified milk, sour milk or yogurt: 2µg

200ml of vitamin D-fortified plant-based products: 2µg

150g of funnel chanterelles (Craterellus tubaeformis): 23µg

FOOD FOR A HEALTHIER SKIN

COLD POTATOES?

I'm not joking — cold potatoes kept in the fridge overnight contain resistant starch, which is hard for the body to break down. Instead, it goes straight through to the large intestine, where it provides food for good bacteria.

VITAMIN E

Vitamin E is an important fat-soluble antioxidant that has been used for over fifty years in medical treatments of skin disorders, as well as in cosmetic skincare. It's been used to treat everything from eczema and acne to yellow nails. Though it's naturally present in our skin, vitamin E is deactivated by UV rays, and vitamin C (or some other antioxidant) is needed to regenerate it. Like vitamin C, it's been shown to have photoprotective properties, meaning that it shields us against UV damage to the skin. Though vitamin E deficiency can accompany certain medical conditions such as inflammatory bowel disease (IBD), it's otherwise unusual.

This is how much vitamin E you need daily:

Women: 8mg
Men: 10mg

You'll find vitamin E in these foods:

50g of sunflower seeds: 7.6mg
50g of almonds: 7mg
Three tbsp of rapeseed oil: 10mg
One egg: 0.6mg
One avocado: 2mg

ZINC ACTIVATES A SUPERENZYME

Zinc, a mineral with many important functions, activates around 100 enzymes that affect the immune system and cell division. These include the powerful antioxidant enzyme *superoxide dismutase*, which reduces oxidative stress in the skin.

The skin contains up to 6 per cent of all the zinc in the body. Levels are highest in the epidermis, which contains about five times as much zinc as the dermis. Zinc is particularly important to the immune system, so it's not surprising that zinc deficiency produces various noticeable symptoms including acne, eczema, slow healing of wounds, dry and sensitive skin, seborrhoeic eczema, and vitiligo.

Animal protein facilitates zinc uptake. Combining nuts with yogurt is a good idea, and so is eating wholemeal bread with a slice of ham.

This is how much zinc you need daily:

Women: 7mg
Men: 9mg

You'll find zinc in these foods:

100g of beef: 4.3mg
200ml of natural yogurt: 3mg
Two slices of wholemeal bread: 1mg
250g of stewed liver: 7.5mg
100g of mussels (cooked): 2.8mg
Two slices of cheese: 1mg
30g of nuts: 1.4mg
100ml of rolled oats: 1mg

SELENIUM — THE MINERAL WITH ANTI-AGEING POTENTIAL

Selenium is an important antioxidant that protects the cells of your body from oxidative stress, one of whose effects is ageing. One of selenium's key functions is activating an antioxidant enzyme called *glutathione peroxidase* in the skin. It's even more effective in conjunction with vitamin E.

In 2017, scientists in Lyon found that selenium protects the keratinocytes in the epidermis from becoming passive and senescent. As a result, it's been proposed that selenium be used to counteract skin ageing, but there haven't yet been any in-depth studies.

A link has also been found between low zinc levels and many of our common skin disorders, including psoriasis, acne, vitiligo, and seborrhoeic eczema. Some experimental studies suggest that selenium deficiency lowers resistance to the *Candida albicans* yeast.

This is how much selenium you need daily:

Women: 40µg
Men: 50µg

You'll find selenium in these foods:

125g of salmon: 42µg
125g of cod: 34µg
100g of marinated herring: 14µg
One egg: 16µg
100g of brown rice (before cooking): 13µg
80g of tinned chickpeas: 16µg
One Brazil nut: 43µg
75 peeled prawns: 18µg

FOOD FOR A HEALTHIER SKIN

Recipe: Sweet and Savoury Nut Mix

Skin Combo: Selenium + Vitamin E

SWEET AND SAVOURY NUT MIX

Selenium: Brazil nuts
Vitamin E: sunflower seeds, almonds
100g of Brazil nuts
100g of almonds
50g of sunflower seeds
3 tbsp of liquid honey
Sea salt
Rosemary

Instructions

Set the oven at 150°C. Chop the Brazil nuts finely. Mix the chopped nuts with the sea salt on a baking tray covered with greaseproof paper. Drizzle with honey and sprinkle with rosemary. Cook in the middle of the oven for fifteen minutes.

Omega 3 and Omega 6 — A Pair of Good Fatty Acids

Studies show that a deficiency of omega 3 and omega 6 polyunsaturated fatty acids massively boosts transepidermal water loss (TEWL). An American study has shown that women with a higher omega 6 intake (in combination with vitamin C) have better skin quality, higher moisture levels, and fewer wrinkles. A link has also been demonstrated between eczema and omega 3 and 6 deficiency.

Fatty acids have an anti-inflammatory effect on the skin, so they're particularly beneficial for people with a skin inflammation such as acne or eczema. There are clinical studies showing that omega 3 supplements can reduce skin inflammation associated with disorders like psoriasis or eczema. An interesting German study showed that women who took flaxseed oil (rich in omega 3) every day for three months had softer, smoother, and better-hydrated skin than those who didn't.

Omega 6 is found in a number of foods that we eat in our daily lives. Inexpensive oils like maize (corn) oil, for example, are rich in omega 6. You'll find this oil in cakes, biscuits, fried onions, fast food, and crisps, foods that aren't good for the skin. Instead, why not eat plenty of nuts and seeds, which supply not only omega 6 but also other good fats, minerals, and fibre?

You'll find omega 3 oils in these foods:

Oily fish such as salmon, mackerel, herring, and sardines; algae; walnuts; rapeseed oil; and flaxseed (linseed) oil

You'll find omega 6 oils in these foods:

Sesame seeds, sunflower seeds, walnuts, pumpkin seeds

ESSENTIAL AMINO ACIDS

Amino acids, the building blocks that make up proteins, are important for all organs, including our skin. That's why it's important to include protein in the diet.

There are twenty amino acids. The nine which the body can't produce itself are known as essential amino acids. These are: phenylalanine, histidine, tryptophan, isoleucine, methionine, threonine, leucine, lysine, and valine.

The so-called non-essential amino acids are produced by the body, but only with the help of the essential amino acids. That means that a deficiency of essential amino acids can also lead to a deficiency of non-essential ones.

One amino acid that's particularly interesting in connection with the skin is histidine, which has been shown to have an impact on dry or eczematous skin. A British study shows that giving histidine supplements to adult eczema patients alleviates eczema as much as applying a mild cortisone cream. It's also been shown that amino acids can boost blood circulation.

You'll find amino acids in these foods:

Since amino acids are the building blocks that make up proteins, you'll find them in protein-rich foods, especially those of animal origin.

Meat, fish, eggs, milk, and dairy produce; soya products, legumes, soya beans, nuts, and grains.

Recipe: Overnight Oats

Skin Combo: Omega 3 + Omega 6 + Vitamin C

Omega 3: walnuts
Omega 6: sunflower seeds
Vitamin C: strawberries

Overnight oats with sunflower seeds, walnuts, and strawberries — one portion

100g of rolled oats
One tbsp of sunflower seeds
200ml of milk (any kind)
One pinch of vanilla powder

Topping

Six sliced strawberries (fresh or frozen)
Half a banana, sliced
Five walnuts, chopped

Instructions

Mix together all the ingredients apart from the topping in a large glass or a glass jar. Put the glass in the fridge overnight. In the morning, stir the mixture and add the topping. Bon appétit!

Prebiotics and Probiotics — Gut Foods That Also Feed Your Skin

In the chapter on the microbiome (p. 126), we looked at the importance for skin health of supporting the gut microbiome. We can do this by consuming polyphenols, but also by making sure we have prebiotics and probiotics in our diet. Research into prebiotics and probiotics is a hot topic right now. There are more and more studies to show that they have a positive impact on skin conditions such as eczema, acne, and rosacea.

Prebiotics are fibres for the micro-organisms living inside us. They can be found in foods like onions, garlic, Jerusalem artichokes, unripe bananas, unripe banana flour, beans, seaweed, rolled oats, and cold boiled potatoes. These provide food for the good bacteria in the gut.

Probiotics are foods containing live bacteria. They can be found in fermented foods including kimchi, sauerkraut, kefir, natural yogurt, kombucha, tempeh, and miso.

ABOUT LIFESTYLE

TIME TO START SNACKING ON SUPPLEMENTS?

Since there's a clear link between many of the nutrients we need and our skin, it's understandable if our thoughts turn to dietary supplements. But beware — unless you're sure you're suffering from a specific deficiency, food supplements are not the best solution. There's always a risk of overdoing it. Some vitamins and minerals compete with each other to be absorbed into the body. Take a lot of iron, and you might risk copper deficiency!

So start by treating yourself to nutritious food and see what it can do for your skin, especially if you have specific skin problems such as dryness, sensitive skin, spots or pimples, pallid skin, or rosacea.

How to Delay Ageing

Changes in the skin are the most visible sign of ageing. The skin's elasticity declines, and it becomes more wrinkled, drier, and paler. Many people get liver spots as well.

How your skin ages depends on a range of factors including your environment, exposure to UV light, your diet, and other aspects of your lifestyle, such as whether you smoke.

Intensive research is under way to identify the best diet for a long life and healthy ageing. Such research often focuses on how to help our cells remain young and postpone senescence.

In one study, carried out in the Netherlands in 2018, scientists looked at the impact of diet on wrinkles. This was a large-scale study involving 2750 Dutch people of both sexes. Women who ate a lot of snacks and red meat were seen to have deeper facial wrinkles than those who ate plenty of fruit. No difference was found in men.

In another large-scale study carried out in Australia, various countries (Sweden, Greece, and Australia itself) were compared. Researchers looked at the impact of diet on the formation of wrinkles and on skin ageing in general. More vegetables, olive oil, and pulses, along with a reduced intake of butter and margarine, meat, milk products, and sugary foods, were found to result in a more youthful appearance.

Japan's small Okinawa archipelago has been much researched in recent years. The reason for this is the inhabitants' longevity, for which they probably have their lifestyle to thank. The diet also seems to be one of the main reasons that Okinawa has the world's highest proportion of centenarians. The Okinawan diet is rich in vegetables, soya-based foods, fish and algae, sweet potatoes, and herbs and spices. Not only do Okinawans have a nutritious diet, they're also sparing with calories; people stop eating when they're 80 per cent full. The Confucian motto that expresses this is *hara hachi bu*. Restricting one's calorie intake has in fact been proven to slow down the ageing process.

HOW TO DELAY AGEING

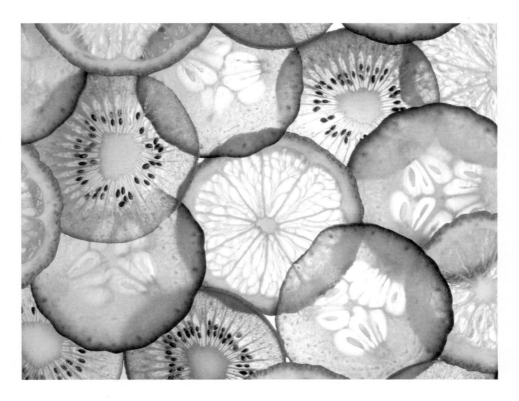

Rejuvenating Polyphenols

Polyphenols are an enormous group of antioxidants; 8000 have been identified so far. They all have a similar chemical structure and are divided up into various classes, three of the most important being stilbenes, tannins, and flavonoids.

Polyphenols are the group of antioxidants that have been most researched in connection with skin ageing. A diet rich in polyphenols is believed to be good for both our internal organs and our external one — the skin.

ABOUT LIFESTYLE

Polyphenols have a very positive effect on the body's micro-organisms. They provide excellent food for these micro-organisms — and bacteria, in their turn, help optimise our uptake of the nutrients from polyphenols. A diet rich in polyphenols has also been shown to balance the bacterial flora in the gut and to raise the levels of good bacteria such as lactobacilli and bifidobacteria. When these bacteria take in polyphenols, they produce short fatty acids, which have an anti-inflammatory effect on the skin.

HYDRATING DRINK BASED ON GREEN COFFEE BEANS

In a study conducted by Japanese scientists in 2017, polyphenols extracted from green coffee beans were shown to improve skin health. The study was carried out on forty-nine female students with dry skin. By the end of the study, the women who had been given the polyphenol-rich drink showed less transepidermal water loss (TEWL) and better-hydrated skin than those who had been given the placebo. The women who had been given the polyphenol-rich drink also had better microcirculation and more free fatty acids in their skin.

STILBENES AND TANNINS

One of the stilbenes often associated with the skin is resveratrol. Though it's long been talked about as a super-antioxidant and an anti-ageing molecule, there aren't many studies to show that it really does have an anti-ageing effect. However, it's been shown to protect against ageing caused by the sun's UV rays.

One study combined resveratrol and pro-cyanidine, a tannin. After participants had taken this combination for sixty days, their skin became significantly more resistant to oxidation. There were also improvements in hydration and elasticity, a decrease in the number of liver spots, and smoother skin texture. It's not clear whether these changes were caused by resveratrol, procyanidine, or the combination of the two substances.

RESVERATROL

CURCUMIN

Recipe: Spicy Polyphenol Bonanza

Skin Combo: Phyto-Oestrogen Resveratrol + Procyanidin

Resveratrol: red grapes (peeled), cocoa beans, blueberries, and pistachios
Procyanidin: cinnamon, red grapes (peeled), cocoa beans, and apples

SPICY POLYPHENOL BONANZA — one portion

Half an apple (unpeeled)
10 organic red grapes
100g of blueberries (fresh or frozen)

Topping

Two tbsp of pistachios
One tsp of raw cocoa (cacao) nibs
A sprinkling of cinnamon
A little liquid honey

Instructions

Chop the apple, cut the grapes in half, and mix with the blueberries. Top with the pistachios and cocoa nibs, and finish with cinnamon and honey.

FLAVONOIDS

One of the thousands of flavonoids is epigallo-catechin gallate (EGCG), best known as the substance responsible for the antioxidant effect of green tea. The use of EGCG has long been discussed as a way of protecting the skin and our body — and even as an antidote to cancer.

Isoflavones are another group of flavonoids with powerful antioxidant properties. They are often called phyto-oestrogens because of the similarity between their chemical structure and that of oestrogen. The soya isoflavone genistein is a well-known anti-ageing substance. There's been some speculation about whether a soya-rich diet containing these isoflavones in countries like China and Japan is responsible for the fact that the inhabitants have firmer skin and fewer wrinkles on average than Europeans. Anthocyanins are another well-known group of flavonoids found in high concentrations in dark-coloured berries, fruit, and vegetables. Blackcurrants, blueberries, and blackberries are extremely rich in anthocyanins. Blackcurrant extract, for example, has been shown to boost the formation of collagen, elastin, and hyaluronic acid in fibroblasts.

Flavonoids known to affect the skin:

resveratrol, procyanidin, genistein, EGCG, anthocyanins, naringenin, quercetin, and curcumin

GENISTEIN

Sources of Polyphenols

A French study pinpointing the 100 foods that are richest in polyphenols puts cloves in first place. Cloves have a polyphenol content of over 15 per cent. A number of other spices have a high ranking. They include dried peppermint, which comes in second with a polyphenol content of 12 per cent, and star anise, which comes in third with a polyphenol content of around 5 per cent. Other spices, such as coriander, basil, ginger, cinnamon, and turmeric also contain polyphenols, but not in such large quantities. Cocoa powder proved to be the fourth richest source of polyphenols, with 3.5 per cent. Dark chocolate came in eighth, with 1.6 per cent.

A number of dark-coloured berries are rich in polyphenols. They include popular and readily available kinds such as blackcurrants, blueberries, blackberries, strawberries, and raspberries. The berry with the highest polyphenol content is the aronia berry (chokeberry), with over 1.7 per cent. You may well find these growing in a hedge near you. Sadly, they don't taste very good.

The skin or peel of many fruits and berries is particularly rich in polyphenols. Examples include citrus fruits, apples, and grapes.

High levels of polyphenols have been found in some nuts (both raw and roasted), such as hazelnuts, pecan nuts, and almonds.

There are also plenty of vegetables containing polyphenols, but they tend to have lower levels than fruits and berries. Vegetables with higher polyphenol levels include artichokes, olives, red onions, broccoli, and spinach.

Among soya-based foods, the following are rich in polyphenols: tempeh, soya bean meal, soya yogurt, and soya beans.

The product most often mentioned in

Senolytics

Substances that inhibit the development of senescent cells are known as senolytics.

These substances are: vitamin E (almonds), EGCG (green tea), quercetin (capers, black radish, watercress, cabbage, cranberries, lingon-berries), genistein (soya), resveratrol (red grapes), silibinin and silymarin (milk thistle or *Silybum marianum*), sulphoraphan (cauliflower, broccoli, Brussels sprouts), naringenin (citrus peel), fisetin (straw-berries, apples, and grapes), floretin (apple peel), and allicin (garlic).

Recently a team of scientists from the US showed that fisetin was the flavonoid with the most powerful senolytic effect.

SENOLYTIC DRINK

Epigallocatechin gallate (EGCG): green tea
Naringenin: citrus peel
Resveratrol: red grapes
Fisetin: strawberries
200ml green tea
Grated peel and flesh of an organic orange
Ten organic red grapes
Six strawberries

Put all the ingredients in a mixer and switch it on full. Pour into a glass. Cheers!

HOW TO DELAY AGEING

connection with polyphenols, however, is red wine. You may have heard how beneficial it is to health in moderate amounts, thanks largely to resveratrol.

THE FRUIT FLY AND THE AYURVEDIC BREW

In August 2013, I had the pleasure of taking part in a conference on ageing. Over 100 scientists from all over the world spent a week in a mountain village above the Italian city of Lucca — where the starring role in the conference fell to the humble fruit fly.

Most research into how we age and what genes are involved is carried out on the little fruit fly, *Drosophila melanogaster*.

It would be tricky to conduct such studies on human beings, as you'd have to wait eighty to 100 years to see the outcome.

The fruit fly study showed that many genes play a role in the ageing process, particularly those that can be activated or switched off through a limited calorie intake. What's interesting is that you can tweak these genes yourself by restricting the amount you eat. Of all the genes that affect how we age, the one most studied is IGF-1, which has been the subject of over 3000 scientific studies. Now, several years after the conference, countless books and articles have been published on that very subject, and more or less half the population has tried out diets based on restricting one's calorie intake, such as the 5:2 diet and the Blue Zones diet.

There are other interesting studies involving the fruit fly. The longevity of fruit flies fed on a combination of probiotics and the Ayurvedic remedy triphala increases by 60 per cent. Instead of living for forty days, they can reach the grand old age of sixty-six days. In these studies, they also showed fewer signs of ageing (such as increased resistance to insulin, inflammation, and oxidative stress).

Polyphenols from sources such as green tea (EGCG) or selenium also make fruit flies live longer. Conversely, sodium glutamate, a common food additive, reduces their lifespan. You might want to bear that in mind when you're eating foods that contain additives. Glutamate, a flavour enhancer, is found in stock cubes, snacks, and ready meals.

SUNSCREEN YOU CAN EAT

Foods rich in carotenoids can make you less sensitive to sunlight! When the sun's UVA and UVB rays reach the skin, they cause oxidative stress. Studies show that carotenoids can mitigate this stress, thereby reducing the skin's light-sensitivity. In a similar study conducted recently in Germany, sixty-five women consumed lutein, a carotenoid, for twelve weeks. This was observed to reduce levels of the enzyme MMP, the baddie when it comes to ageing. In other words, lutein inhibits sun-related ageing.

It's not just carotenoids that make people less sensitive to the sun. Another substance that inhibits the sun's harmful effects is vitamin D. Clinical studies have shown that people who take vitamin D have less sun-related inflammation than those who don't.

How about chocolate as a sunscreen? Research conducted in 2009 on thirty healthy participants showed that raw chocolate, which is rich in flavonoids, can boost your resistance to the sun's rays. Subjects who ate 20g of raw chocolate a day for three months were able to spend twice as long in the sun as those who ate the same quantity of ordinary dark chocolate. Making chocolate without exposing the ingredients to oven temperatures preserves the flavonoids in the chocolate, which probably makes all the difference.

Fish oil as a sunscreen? In a three-month study with forty-two participants, those who took omega 3 four times a day were found to tolerate the sun better. They had less skin irritation and there was less damage to their DNA. A follow-up study showed that ingesting 10g of fish oil a day produced the same results. Participants' skin became twice as resistant to UV rays.

Vitamins C and E plus carotenoids as a sunscreen? A number of studies have shown that a mix of vitamins C and E can have a big impact on skin ageing. A mix of carotenoids and vitamins E and C has also been shown to be effective in counteracting sun-induced signs of ageing.

Can these substances reduce the need for shade, protective clothing, or protective sun cream? The answer's no. It's important to stress that fish oil, carrots or vitamin D can't provide full protection against the sun. It might well be a good idea to consume more fish oil, and carrots during the six warmest months of the year — and maybe this is a good reason to eat chocolate! And why not try our sun smoothie? Despite these benefits, however, I don't recommend trying to replace shade, clothes, or sun cream with the food you eat.

Recipe: Sun Smoothie

Skin Combo: Vitamin C + Vitamin E + Carotenoids

Vitamin C: strawberries, citrus fruits
Vitamin E: rapeseed oil
Carotenoids: carrots, mango

SUN SMOOTHIE

One finely grated carrot
100g of frozen mango
Ten strawberries (fresh or frozen)
One lemon
Three centimetres of finely grated ginger
One tbsp of cold pressed rapeseed oil
200ml of good-quality apple juice

Instructions

Put all the ingredients in a mixer and
switch on full.

HOW TO DELAY AGEING

The Western Diet

Processed food, red meat, dairy products, trans fats, empty carbohydrates, and low-fibre food — these are typical of the Western diet.

Increasing Westernisation in more and more Eastern cultures has gone hand in hand with a rise in skin complaints such as acne.

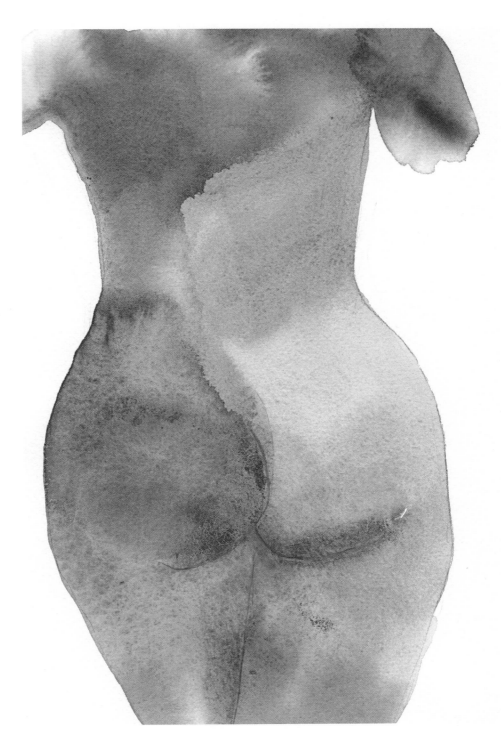

Acne is linked with what we eat. In fact, it's been linked specifically with the Western diet. Eighty per cent of all young people in the West suffer to some extent from acne, and although for most people it's a typical puberty-related problem, it's striking how many adults long past their teens suffer from the condition. Acne has exploded globally — so much so that the word 'epidemic' is hardly an exaggeration.

Dermatologist Kathleen Woolf and her team conducted an interesting study of the factors conducive to acne in young New Yorkers. Participants with moderate to serious acne were seen to consume more carbohydrates than acne-free participants. Those with moderate to serious acne also had higher insulin levels, concentrations of insulin-like growth factor-1 (IGF-1), and higher insulin resistance than non-acne-sufferers.

In certain areas of New Guinea, Paraguay, and Brazil, where people have a more Palaeolithic diet — dairy-free and with a low glycaemic load — they are spared acne altogether. There are entire acne-free populations. Conversely, a rise in the incidence of acne has been reported among the people of Japan's Okinawa islands and in Chinese people, as a result of abandoning their traditional diet and adopting one with more soft drinks and hamburgers.

As regards acne and many other diseases of affluence, science has now discovered another villain — trans fatty acids. These cause acne by accelerating sebum production and cornification in the pores, which then become rapidly blocked.

You'll find trans fats in foods whose overall nutritional value is low, such as fast food, crisps, chips (French fries), pre-packed pizzas, and mass-produced cakes. To find out whether a product contains trans fats, check the list of ingredients. If they include 'partially hydrogenated fat', that means trans fat. 'Hydrogenated' or 'fully hydrogenated' fats, on the other hand, are not trans fats.

Trans fatty acids occur in very small quantities in nature, in the stomachs of ruminants. That's why there are small amounts in milk and milk products, in fat from ruminants, and, as a result, in sausages and certain meat products.

IS THE SKIN AFFECTED BY SUGAR?

Perhaps unsurprisingly, our skin isn't fond of sugar, an energy-rich but nutrient-poor substance. This is because the sugar molecule readily reacts with proteins and fats in the body, resulting in what are known as advanced glycation end-products (AGEs for short). In other words, AGEs are glycated (sweetened) molecules. Collagen, a protein, is vulnerable to the damage caused by the sugar molecule, and research has shown that as much as half the collagen in the skin can undergo glycation during a lifetime.

So what's the problem with glycation? Well, it's thought to reduce skin elasticity. The formation of AGEs speeds up the ageing process and reinforces other signs of age, such as increased pigmentation and a sallower skin tone. The level of AGEs is also a marker in sun-damaged skin. The increase in AGEs in skin that has been extensively exposed to the sun is believed to be a result of UV radiation, which heightens oxidative stress and thereby speeds up the process of AGE formation. Skin that has been protected against the sun shows virtually no sign of AGEs.

The first reports about how sugar levels in both the blood and the skin can be related to a sugar-rich diet were published a good seventy years ago. Although the significance of these studies and their results were not fully understood at the time, they nonetheless revealed an intriguing link between diet and skin health. We now understand that food is a source of monosaccharides, which, if present in large quantities, result in AGEs in the body and the skin.

AGEs are also produced by certain cooking processes. Grilling, frying, and roasting are known to produce relatively high levels of AGEs. In contrast, water-based cooking methods such as boiling and steaming produce fewer AGEs. The types of diet that have been shown to limit the formation of AGEs most are the Mediterranean diet and other diets with a low glycaemic index (GI), coupled with the consumption of polyphenols. Studies have also shown that AGEs can be reduced by consuming ginger, garlic, various antioxidants such as lipoic acid (found in spinach, broccoli, potatoes, tomatoes, Brussels sprouts, carrots, and beetroot) or carnitine (meat, chicken, milk, cheese, and asparagus), or flavonoids (such as EGCG), vitamin E, selenium, zinc, and manganese.

DOES MILK GIVE YOU PIMPLES?

When I was growing up my dad was always anti-milk. 'We're not calves, are we?' he'd say. My mum, on the other hand, would pour out a big glass of milk for me and my brother. We'd have it together with meatballs, potatoes, and a creamy sauce. Mum herself would take a packet of Lean Cuisine out of the freezer, and Dad would have some beetroot juice or a glass of red wine.

Dad had a point, at least as far as milk was concerned. Milk is no ordinary food, but one with a unique biological function. It stimulates a number of growth hormones, one of which is IGF-1, which makes new-born humans and animals grow. Growth is complete once we reach adulthood, which is why some scientists think it's a bad thing to consume a lot of milk at this later stage of our lives. Yet milk is an invaluable source of essential amino acids, and it contains many vitamins and minerals.

If you don't have acne and you're partial to milk or cheese, what should you do with this information?

Both milk and cheese are nutritious foods — so if your skin doesn't react to them, there's no need to make life harder for yourself and change your diet just for the sake of it. However, if you're prone to acne, it may well be a good idea to try cutting down the milk and dairy products in your diet. Other skin conditions which may improve thanks to reducing or excluding dairy products from the diet are psoriasis and eczema (atopic dermatitis). But this isn't a clear-cut issue. Some studies have shown that drinking milk improves the condition of children suffering from eczema. But this conclusion applies only to fresh milk, which suggests that the significant factor may be the lactobacilli in unpasteurised milk.

Other Skin Disorders That Are Affected by What We Eat

ECZEMA

A 2017 study showed that cutting out certain types of food had a positive effect on eczema. Most improvement was reported in those who ate no products containing white flour, gluten, or vegetables belonging to the potato family (potatoes themselves, tomatoes, egg-plants, and capsicums). Participants also reported that eating vegetables, fish oil, and fruit improved the condition of their skin.

PSORIASIS

In a major study conducted in Denmark in 2018, a 75 per cent improvement in psoriasis plaques was observed in patients who, in addition to undergoing treatment, also included more fibre in their diet and cut down on red and processed meat. In another study carried out in San Francisco in 2017 on over a thousand psoriasis sufferers, participants experienced a significant improvement when they reduced their consumption of alcohol, gluten, and vegetables from the potato family while also increasing their intake of fish oil or omega 3, vegetables, and vitamin D.

Participants who were already on a vegan or Palaeolithic diet reported particularly positive outcomes. The 'Palaeolithic diet', based on Stone Age foods, involves cutting out foods such as dairy products, refined sugar, processed foods, beans, lentils, and potatoes. Following this diet means eating fruit and vegetables, nuts and seeds, lean meat (especially game), fish rich in omega 3, and oils obtained from fruits and nuts, such as olive oil.

ROSACEA

Relatively little research has been done into rosacea and the diet by comparison with other skin disorders. In one of the few studies that exists (carried out by the US National Rosacea Society), 78 per cent of participants changed their diet, and 95 per cent of this group reported significant improvements, especially from cutting out spicy sauces, cayenne pepper, and chili pepper. Many people also reported improvements thanks to cutting out tomatoes, chocolate, alcohol, coffee, tea, and citrus fruits.

It is the capsaicin and cinnamaldehyde in these foods that trigger the TRPV1 receptor in the skin, stimulating blood vessels so that they open, resulting in a sensation of heat, flushing, and irritation.

There's no convincing proof that any specific substances in food can relieve the symptoms of rosacea. However, promising results from a small number of studies suggest that omega 3 supplements may have some effect. Studies of how zinc supplements affect rosacea have produced contradictory results. While one study found a significant improvement in people taking 100mg of zinc sulphate three times a day, another one found no improvement in participants who had taken 220mg of zinc sulphate twice daily for ninety days.

THE WESTERN DIET

PSORIASIS

In a major study conducted in Denmark in 2018, a 75 per cent improvement in psoriasis plaques was observed in patients who, in addition to undergoing treatment, also included more fibre in their diet and cut down on red and processed meat. In another study carried out in San Francisco in 2017 on over a thousand psoriasis sufferers, participants experienced a significant improvement when they reduced their consumption of alcohol, gluten, and vegetables from the potato family while also increasing their intake of fish oil or omega 3, vegetables, and vitamin D.

Participants who were already on a vegan or Palaeolithic diet reported particularly positive outcomes. The 'Palaeolithic diet', based on Stone Age foods, involves cutting out foods such as dairy products, refined sugar, processed foods, beans, lentils, and potatoes. Following this diet means eating fruit and vegetables, nuts and seeds, lean meat (especially game), fish rich in omega 3, and oils obtained from fruits and nuts, such as olive oil.

CHRONIC GASTROINTESTINAL DISEASES

Gastrointestinal diseases that affect sufferers' skin are an intriguing illustration of the gut–skin axis. People with Crohn's disease, inflammatory bowel disease (IBD), or irritable bowel syndrome (IBS) often suffer from inflammatory skin disorders such as atopic eczema, dermatitis, mouth ulcers, or psoriasis. Gastrointestinal inflammation in these patients can be transferred to the skin, leading to skin inflammation, and on top of that, they have skin problems arising from the body's inability to absorb all the nutrients it needs.

DERMATITIS HERPETIFORMIS (DH)

Although most people can tolerate wheat, it is nutrient-poor, and extensive consumption of wheat brings with it a risk of nutrient deficiency. Those who can't eat wheat are either allergic to it or gluten-intolerant. If you have gluten intolerance, you won't be able to eat rye, maize, or oats either, even though oats contain no gluten. Oats go through the same mills as other cereals, |so there is a risk that they may come into contact with small amounts of gluten.

Gluten intolerance is an autoimmune disease affecting 1–2 per cent of the population. Their immune system reacts against gliadin and glutenin, proteins found in wheat, rye, and maize. The skin condition associated with gluten intolerance, dermatitis herpetiformis (DH), manifests as an itchy rash and blisters. It takes several months for a gluten-free diet to have a visible impact on the skin. In such cases, a gluten-free diet is a must.

The skin conditions linked less specifically with gluten sensitivity are hives (urticaria or nettle rash), atopic dermatitis, psoriasis, mouth blisters, vitiligo, and rosacea.

People suffering from these disorders may sometimes experience an improvement in their condition if they switch to a gluten-free diet.

THE WESTERN DIET

Feeding Your Skin — A Guide to Nutrition

Your skin loves nutrients. They're transported via the bloodstream to your skin, where they can really make a difference. I wouldn't go as far as to claim that carrots will get rid of any crows' feet around your eyes, or that a portion of kimchi will cure acne or dry skin — but it's as clear as day that nutritious food is good for your skin. And skin that feels good looks healthier.

So what should you eat to have a healthy skin? The best tip we can give is to start with the most researched diet ever — the Mediterranean diet.

Mediterranean food, which includes large amounts of fruit and vegetables, fish and shellfish, pulses, wholegrains, seeds and nuts, yogurt, cheese, and olive oil — with a little red wine — is good not only for the heart, gut, and health in general, but also, specifically, for the skin.

The different types of fibre in which Mediterranean food is rich optimise the gut's capacity to boost skin health, while yogurt gives us probiotics, and fish contribute omega 3 oils.

Mediterranean food is also rich in polyphenols and carotenoids — antioxidant 'superheroes' that are particularly beneficial to the skin.

Scientists have shown that people who eat a Mediterranean diet have higher concentrations of carotenoids in the blood compared with those who don't — and carotenoids are known to be good for the skin. The Mediterranean diet has also been shown to be effective in inhibiting the formation of glycated (sweetened) proteins and fats (AGEs).

If the way you eat already resembles the Mediterranean diet, you're already well on the way to a healthier skin!

Skin Foods — A Nutrition Guide

You'll automatically get the nutrients you need by eating a Mediterranean diet, but there are a few tips you can follow to give your skin an extra boost.

By 'skin foods', we mean foods that are particularly rich in nutrients and scientifically proven to be especially good for the skin. Use this guide to enrich your diet with particularly skin-friendly nutrients.

TWO PS

Low-carb diets are in vogue at the moment. But if you're on a strict low-carb diet, bear in mind that it won't give you much of the prebiotic fibre that makes for a healthy gut and, by extension, a healthy skin. It may also be a good idea to make sure that your carbohydrates come from fibre-rich sources such as lentils, beans, oats, or potatoes, not from low-fibre white bread or pasta. In this book we've stressed the importance of boosting your gut flora to improve skin health, and consciously eating prebiotics and probiotics is a way to strengthen your skin from within. Remember the South Koreans who consumed the galacto-oligosaccharide found in lentils, beans and seaweed for three months, which gave them better-hydrated skin and fewer wrinkles? Have another look at p. 129 in 'About the Microbiome'.

PREBIOTICS

We consume prebiotics in the form of various kinds of fibre and complex carbohydrates, found in particularly large quantities in onions, garlic, oats, beans and lentils, Jerusalem artichokes, unripe bananas, and cold boiled potatoes. High levels of resistant starch, a prebiotic fibre, are found in potato flour, and this has induced uncon-ditional fans of the gut microbiome to take one to two tablespoonsful of potato flour in cold water, like a shot. However, increasing the amount of fibre you eat can give your guts more than they bargained for if you're not used to that kind of food, so do exercise caution when boosting your intake.

PROBIOTICS

There are various ways to take in the good micro-organisms found in probiotics. A wide range of lactic acid bacteria are available in tablet, capsule, or liquid form. Examples include *L. reuteri*, *L. rhamnosus*, *L. paracasei*, *L. plantarum*, *L. acidophilus* and *L. fermentum*. However, you can also add lactic acid bacteria to your diet by eating fermented vegetables such as sauerkraut, kimchi, or kombucha. Yogurt, sour milk, and cheese also contain good lactobacilli. Choose unpasteurised or bacteria-enriched products, as pasteurisation destroys lactic acid bacteria. Remember the Japanese students who drank curdled milk (filmjölk) enriched with *L. breve* for a month, which gave them better-hydrated skin and fewer wrinkles! (See p. 131 in 'About the Microbiome'.) Probiotics have also proven to be particularly beneficial in treating eczema and acne.

FEEDING YOUR SKIN — A GUIDE TO NUTRITION

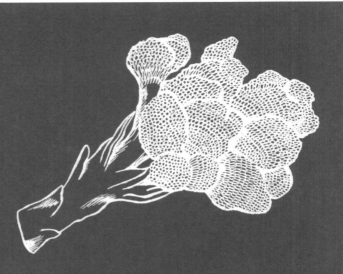

Vegetables

For those living in northern Europe, carrots are an excellent source of beta-carotene, a substance which we need to stimulate collagen formation. I eat a bag of snack-sized carrots or two large carrots a day. Enrich your diet with other carotenoids as well, such as lycopenes (found in tomatoes) or astaxanthin (found in salmon and microalgae).

Why not add a spoonful of algae such as chlorella to your morning smoothie? And try out the *carotenoid-rich gazpacho* recipe on p. 144.

Vitamins A, B, C, D, and E are especially important for the skin. Spinach and curly kale are important sources of folate and vitamin C. Spinach also contains lipoic acid, an antioxidant. Try frying it in rapeseed oil with plenty of garlic and a dash of salt and pepper!

CARROTS	Beta-carotene and lipoic acid, which limit the incidence of AGEs
SPINACH	Folate, vitamin C, and lipoic acid
CURLY KALE	Folate and vitamin C
BROCCOLI	Lipoic acid, senolytic sulphorophane
MICROALGAE	Astaxanthin
COLD POTATOES	Prebiotics
JERUSALEM ARTICHOKES	Prebiotics
FERMENTED VEGETABLES	Prebiotics and vitamin B12
TOMATOES	Lycopene, a carotenoid

Herbs and Spices

Polyphenols are the other important group of antioxidants. They are good for your skin, both as super-antioxidants and as prebiotics that boost your gut flora. Make sure you eat fruit, berries, and vegetables daily, bearing in mind that the skin or peel is particularly rich in polyphenols, and include the following in your diet: nuts, cloves, turmeric, peppermint, cocoa beans, olives and olive oil, soya beans, green tea, and lightly roasted coffee beans. Coffee beans lose much of their polyphenol content if roasted at too high a temperature.

CLOVES	Number 1 among polyphenol-rich foods
TURMERIC	Contains curcumin, a senolytic polyphenol
PEPPERMINT	Number 2 among polyphenol-rich foods
CINNAMON	Polyphenols
GARLIC	Prebiotics and allicin, a senolytic antioxidant
COCOA	Number 3 among polyphenol-rich foods
GINGER	Contains the polyphenol gingerol

Fruits and Berries

Resveratrol is one of the most popular polyphenols — it's often used as a pretext for drinking red wine. Sadly, alcohol isn't good either for your body or your skin, so don't overdo things! A better alternative is raw blueberry juice, as blueberries also contain plenty of resveratrol. And why not try our *spicy polyphenol bonanza* (see p. 157)?

BLUEBERRIES	Vitamin C, polyphenols such as resveratrol and anthocyanins
ORGANIC RED GRAPES	Vitamin C, polyphenols such as resveratrol
APPLES	Vitamin C, fisetin, and phloretin (polyphenols)
LINGONBERRIES	Vitamin C, quercetin
UNRIPE BANANAS	Prebiotics, vitamin B
BLACKCURRANTS	Polyphenols (particularly anthocyanins), vitamin C
BLUE PLUMS	Polyphenols (particularly anthocyanins)
ORGANIC CITRUS FRUITS	Vitamin C and naringenin (in the peel)

Combinations

Salmon is again a good source — of both vitamin B12 and niacin. It's also a good source of omega 3, which helps make your skin softer and reduces dryness. Alternate salmon with mackerel or herring, both of which are rich in omega 3, or opt for a vegetarian omega 3 alternative instead (two tablespoonfuls of rapeseed oil or a handful of walnuts every day).

Many nutrients — selenium and vitamin E, for instance — are more effective if you combine them. Try our sweet-savoury mix (see p. 149), or have a shrimp and avocado salad — both are good ways to make sure you take in selenium and vitamin E together.

Zinc, a mineral, activates superoxide dismutase, a superenzyme that's important to counteract oxidative stress in the skin. Two sources are mussels and oats. Moules marinières, made with an oat-based cream substitute, is a perfect combination!

Our prebiotic *overnight oats* (see p. 153) will provide you with omega 3 and 6 and vitamin C.

BEANS AND LENTILS	Polyphenols and prebiotics
SALMON	Omega 3, astaxanthin (a carotenoid), vitamins B3, B12, and D, and selenium
SHELLFISH	Astaxanthin, selenium, and zinc
NUTS	Polyphenols, omega 3 (walnuts), omega 6, zinc, selenium (Brazil nuts), vitamin E (almonds)
RAPESEED OIL	Omega 3
OLIVE OIL	Polyphenols
SOYA BEANS	Polyphenols, particularly genistein (a flavonoid)
OATS	Prebiotics and zinc

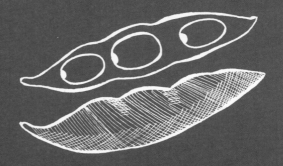

Good and Bad Habits

A warm hug, an affectionate pat on the cheek, or a comforting hand on the shoulder — these are things most of us long for from time to time, perhaps even more so when we're anxious or life seems tough. By massaging someone's tense neck or tired feet for a moment, we can give them a physical boost and a sense of wellbeing.

But touch isn't just a momentary source of comfort and pleasure. In fact, it's been shown to have a wide range of health benefits: it lowers blood pressure and brings the pulse rate down, relieves pain, boosts the immune system, and makes us feel less stressed and more sociable. We can hunger for touch just as we hunger for food — hence the expression 'skin hunger'. Touch makes us feel good because it releases oxytocin, the wellbeing hormone. Oxytocin is classed as being as essential for life as food and water.

Stress

In studies of how stress affects the skin, it has been observed that people suffering from acute stress have less moisture in the skin. Studies of people going through a divorce and of students with exam stress have also shown that the skin heals more slowly under these circumstances, owing to a defect in the barrier function.

Other studies show that people suffering from stress get more spots. And many skin disorders, including vitiligo, psoriasis, and eczema, are aggravated by stress.

Mindfulness is often recommended as the perfect antidote to stress, and there are studies of the impact it can have on the skin. In fact, one study even shows that mindfulness can help wounds heal better. In this study, lesions were made in participants' arms, after which they were divided into two groups. The half who took part in mindfulness therapy for eight weeks healed better than the half who received no therapy.

Some scientists even advocate mindfulness as a therapy for patients with anything from acne to psoriasis and vitiligo. I think that needs to be taken with a pinch of salt. Stress isn't good for the brain, the stomach or the skin — but it's a long way from that insight to the conviction that mindfulness can be used to treat an autoimmune illness such as psoriasis.

Happiness

Unlike people going through a divorce, people in love have higher levels of oxytocin, a chemical secreted when the skin is touched, which lifts mood. Feeling good can make you glow with health — and that applies at any age.

When I was working with vitiligo patients at the Dead Sea centre, we'd sometimes see couples meet and fall in love over their three-week stay, helped by staying at a hotel in a sunny place and enjoying barbecue evenings, Jordanian cuisine, and delicious fresh lemonade with mint. Repigmentation in these couples was very noticeably better than in other people.

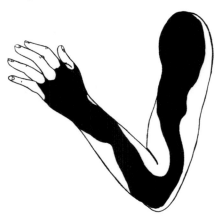

Exercise

Lifestyle magazines often tell us how wonderful exercise is for physique and psyche alike. But what about our largest organ? Is exercise good for our skin as well?

Highly mobile animals have been observed to have healthier fur with less grey in it than more sedentary ones. If exercise can prevent an animal's coat from changing with age, maybe it can do the same for our skin.

In Canada, skin scientists brought together about thirty volunteers of both sexes, aged between twenty and eighty-four, to study the impact of exercise. About half the participants were physically active, engaging in at least three hours of moderate or strenuous physical activity a week, while the rest were sedentary, spending less than an hour a week exercising. As the scientists wanted to look at skin that hadn't had much sun exposure, they opted for taking samples from the volunteers' buttocks. Samples from all participants were examined under the microscope.

When the study was completed, men and women over forty belonging to the half of the group that exercised frequently were found to have a noticeably thinner and healthier stratum corneum, while both epidermis and dermis were thicker and denser. Their skin was more like that of twenty- and thirty-year-olds.

A recent study from Japan also revealed a link between physical activity and fewer advanced glycation end-products (AGEs) in the skin. The implication is that regular exercise makes your skin look younger. However, the mechanisms underpinning this result remain unclear.

Sleep

'Make sure you get your beauty sleep! While you're asleep, your skin is replenishing its moisture levels and regaining its elasticity.'

That's a tip you'll often find in beauty magazines. But is there really anything in it? Well, yes, there is. The hormone which makes our skin healthier through sleep is melatonin. Melatonin is a multi-faceted hormone — but above all it's an effective sleep hormone. It makes us tired and helps us fall asleep more easily.

It also works as an antioxidant, and it's one of the factors that counteract ageing. Melatonin levels rise in the evening as darkness falls. While we're asleep, melatonin is busy repairing damage to our skin, such as that caused by the sun. It also boosts the immune system. People who sleep too little or who sleep in a room that's too light often have low melatonin levels in the bloodstream. This condition can lead to listlessness, fatigue, or even depression. In Sweden melatonin is classed as a medicine, as it is a hormone, and it is available on prescription only.

Outside Sweden, however, melatonin can be bought over the counter as a remedy for everything from jetlag to low mood. It is also sold in skin creams that are claimed to have an anti-ageing effect. However, few clinical studies show it has any such impact.

In a small-scale Italian study of just over thirty women aged over fifty-five, melatonin cream was applied to one side of each participant's face over a three-month period. The skin on the melatonin side was found to be better hydrated, to have fewer wrinkles, and to be both firmer and smoother.

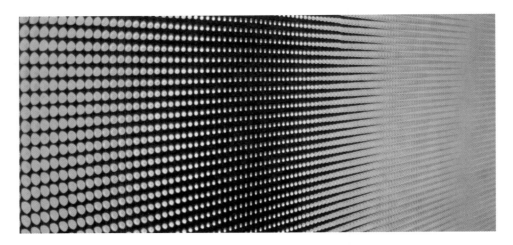

Water

I don't know how many times I've heard people say that dry skin comes from failing to drink enough water. But the idea that water keeps the skin hydrated and youthful is nothing but a persistent myth.

By 2018, there were no fewer than 216 conference articles and twenty-three published studies on the impact of water on the skin. Yet the results are, overall, quite unclear. If you're seriously dehydrated, your skin hydration level will naturally be affected by drinking water — but only under those specific circumstances. Clearly, becoming dehydrated to this degree is dangerous, and fortunately it's rare. A possible exception is vulnerable older people, who are at more risk of becoming seriously dehydrated. But merely 'forgetting to drink enough' during the day will certainly not leave you suffering from skin dehydration.

A daily water intake of one to two litres, as often recommended, is quite adequate. You'll get half of this amount from the food you eat, so it's enough to drink about a litre.

Screen Time

Can blue light from screens speed up the ageing process? There are products on the market now which supposedly protect you against blue light. However, there is no proof that it causes wrinkles. What has been observed is that blue light can cause hyper-pigmentation in people whose skin is darker than skin type 3 (see p. 16–17). It has also been shown to break down the important carotenoids in the skin.

Can you protect yourself by applying ordinary sunscreen? Yes, studies have shown that physical sunscreen affords some protection against blue light. However, we have to look at this realistically. Compare how your skin feels if you sit out in the sun all day with how it feels after the same amount of time at a computer. The difference in strength and impact is obvious. Bearing in mind that blue light has also been claimed to damage the eyes, it's better to get a light filter for your computer, tablet, and mobile than to try to protect your skin by applying sunscreen. Having said this, not nearly enough studies have been carried out yet into blue light and the skin.

ABOUT LIFESTYLE

Smoking

When my team and I were trying to cause oxidative stress in skin cells, we used cigarette smoke or diesel particles for the purpose. This is because smoking forms free radicals and causes inflammation in the skin. We also know it speeds up the skin's ageing process.

Clinical studies have shown that heavy smoking results in premature wrinkles. You might think those wrinkles form only around your mouth, where the cigarette goes, but no — it's been shown that smokers even develop more wrinkles on their arms.

Smoking also makes smokers develop elastosis, meaning that the elastin fibres in the skin lose their elasticity. Their skin becomes wrinkled, sallow, and less firm. Male smokers develop both elastosis and telangiectasia (visible blood vessels near the surface of the skin, also known as spider veins). Female smokers are not affected to the same extent by telangiectasia.

Pollution

There's a great deal of research into how exhaust fumes and pollution affect the skin. The results leave no doubts about their negative impact. The World Health Organization, WHO, published a report in 2016 according to which over three million people die annually from air pollution, making atmospheric pollution the world's number one environmental health risk. Airborne particles are one of the main components of this pollution, and there are increasingly clear signs that they have an adverse impact on human skin; they aggravate inflammation and enzyme activity in MMPs (matrix metallo-peptidases), thereby reducing the amount of collagen and elastin in the skin.

One of the world's biggest cosmetics firms has invested massively in research into air pollution, particularly in China, as more extensive pollution has been recorded there. The skin status of people living in areas with high pollution levels has been compared with that of rural people from areas with cleaner air. People living in the polluted areas have more wrinkles, more patches of pigment, and drier skin.

GOOD AND BAD HABITS

AFTERWORD

Ten Commandments for a Healthy Skin

1. Think *less is more*! A long list of ingredients doesn't mean a better product — in fact, the opposite is true.

2. In relation to the skin, 'natural' means substances that occur naturally in the skin. Look for these in the ingredients list.

3. Pay close attention to molecule size. To have any effect, active ingredients must be able to penetrate the skin.

4. Don't 'over-treat' healthy skin. Remember that not everyone needs skincare. Using more products won't give you healthier skin.

5. Be sparing with skin cleansers. Your skin is smart, and it forms a perfect combination of fats and moisturisers to keep itself soft, pliable, and healthy. Don't wash that layer off for no good reason.

6. Apply sunscreen scrupulously when the UV index is high.

7. Live a healthy life. Take exercise and enjoy nature, fresh air, and being together with other people. Sleep well.

8. Treat your skin's microbiome with respect. Avoid disrupting your micro-organisms, and focus instead on supporting your good bacteria.

9. Eat more carrots!

10. Learn more! The skin is our largest organ and it needs to be given careful consideration. Our knowledge of the skin is developing all the time. Check out the latest articles on *Pubmed.com* or follow us at *Skinomeproject.com*.

LOOKING AFTER YOUR SKIN: THE TEN COMMANDMENTS

AFTERWORD

Time for Change

My fascination with skin began in a Bradford lab nearly two decades ago. I recall how the hairs on my forearms stood on end as, scrutinising my own skin cells under the microscope, I saw the dark pigment beginning to return to the white patches on my skin. There and then, I knew I wanted to devote my professional life to skin research.

And that's just what happened. This book is underpinned by fifteen years' experience of academic dermatology research and the skincare sector — there was just no way I *couldn't* write it. I could see there was a huge amount of interest in how people could make their skin look as good as possible. At the same time, I could see that the level of people's knowledge about how to look after their skin was distinctly less high.
As for the level of public knowledge about the products we spend so much time and money on applying to our skin, that was *very* low.

In writing this book, I've made use of 368 scientific articles from the leading edge of skincare research.

Knowledge in this field continues to develop, and it fascinates me more than ever.

I hope *The Scandinavian Skincare Bible* will at least partially meet the need I've identified by channelling the results of this research to anyone who takes an interest in skin health. Because knowledge changes and develops. And if you know which ingredients to look out for (and which ones to be wary of), it's easier to make the right choices when you're checking out the skincare products on offer. Similarly, knowing what food is good (or bad) for your skin makes it easier to eat healthily.

Over the last few years as a skin scientist, I've taken a close interest in the skin microbiome, and I've appreciated having the chance to focus on it even more while working on this book. It's absolutely fascinating to see how the skin is affected by the gut microbiome and how skin bacteria interact and contribute to good skin health. Research in this field is still in its infancy, but I'd be surprised if, a decade from now, we're still using skincare products that kill the bacteria on our skin — our skin microbiome — as we do today.

Since *The Scandinavian Skincare Bible* takes a close look at skincare ingredients — what works, what doesn't, and what can actually harm the skin — it inevitably involves some criticism of the skincare industry. As I've also worked in this sector, this criticism obviously includes me. However, I'd like to stress that I don't want us to stop using skincare products and start smoothing rainwater into our faces instead. What I *am* hoping for is an improvement in the skincare sector like the one we've seen in the food industry over the last few years. Since additives and processed food first became hot topics, we've learned to ask questions, to be discerning, and to read what's on the packaging. All this has helped to expand the range of good food on the market. I hope the skincare market will go the same way.

The skincare industry needs more innovation — not just in terms of new, effective ingredients, but also as regards packaging, preservation methods, and logistics. If only we could finally replace lidded jars — which we open and close hundreds of times and stick our fingers into — by more product- and consumer-friendly alternatives. If only manufacturers could start printing the date of manufacture on the packaging, so we know whether the product's been in the shop for a month — or several years. These issues are very significant in terms of the product's content and impact. If only there were fewer (or no) preservatives and masking ingredients on the INCI list. If only the rules in place were to change — maybe challenge testing as we know it would be a thing of the past? Just think how good all this would be for our skin.

Now this book's finished, I realise that I need to see it as a first step. There are so many new findings in the skincare field that I want to keep working on.

While I was writing the book, I created a website to share information and promote skin wellbeing. New advances are being made all the time in skin and skincare research. Instead of writing *The Scandinavian Skincare Bible — Part II* straightaway, I'm going to gather together all the information I find at www. skinomeproject.com. In cooperation with like-minded scientists, I hope to be able to pursue more research into what's good for the skin and its microbiome.

It's my ambition to make knowledge of the skin more readily accessible, so that we researchers, users, and producers can put that knowledge to good use — together.

Acknowledgements

Karin Schallreuter

Karin is an emeritus professor at the University of Bradford, a dermatologist, and the founder of the Institute for Pigmentary Disorders in Greifswald, Germany. If Karin hadn't replied when I emailed her in 2002 to ask about a job on her team, I probably wouldn't even be working in dermatology today. During my years in Bradford and Jordan I learned more, and was inspired more, than at any other time in my life. Karin is a real role model to me. Not only has she developed a strategy for treating vitiligo using PC-KUS (pseudocatalase) cream, she has also published nearly 200 scientific articles on vitiligo and malignant melanoma. Karin and her late husband, Professor John Wood, who, sadly, died in 2008, were pioneers in melanogenesis (the study of how pigment is synthesised in the skin), oxidative stress, and vitiligo.

Karin Magnusson

Karin is a registered dietician working in the field of food and communication. She has had a long career in sports nutrition, focusing on elite athletes and the diet they need to optimise performance. As a cognitive behavioural therapist, she has also worked with people like me and you who need help with losing or gaining weight, or who have psychological problems

related to food. From an early stage she began helping her clients to engage in mindful eating, which has only recently become a mainstream concept. When Karin and I began discussing the idea that led to *The Scandinavian Skincare Bible*, we both realised it was essential to include a chapter on the relationship between food and the skin. In the course of her work, Karin has met clients suffering from skin problems who are desperate to find a diet that will help them. The two of us worked together to produce the nutrition guide and the information about skin food set out in the chapter 'About Lifestyle'.

Ina Schuppe Koistinen

Ina Schuppe Koistinen is a co-founder of the Centre for Translational Microbiome Research (CTMR) at the Karolinska Institute, Solna, and a lecturer in the Institute's Department of Physiology and Pharmacology. She has had a long career in the pharmaceutical industry. An expert on the vaginal microbiome, Ina conducts research into how the balance of bacterial flora affects a woman's chances of conceiving and how the vaginal microbiome can be transferred to babies born by C-section. When I met her at a conference on the microbiome in Rotterdam, I asked whether she might like to contribute some watercolour illustrations to this book. She agreed without a second's hesitation. I'm

ACKNOWLEDGEMENTS

so glad now that I went to that get-together in Rotterdam. Ina's watercolours are absolutely lovely, and I'd like to say a big thank you!

Lars Engstrand

Lars is a professor and a co-founder of the Centre for Translational Microbiome Research (CTMR) at the Karolinska Institute. A worldwide expert on the human microbiome, particularly the gut microbiome, he has gone from eliminating pathogenic bacteria during his time at the Swedish Institute for Infectious Disease Control (now the Public Health Agency of Sweden) to working with the body's good bacteria. Lars is a fine researcher and human being whose drive and commitment are truly admirable. In the last few years I have had many fascinating conversations with him, particularly about the skin biome — and he has been a great source of inspiration for this book.

Torborg Hoppe

Torborg is a senior lecturer in the dermatology and venereology section of the Department of Medical Sciences at the University of Uppsala, and a medical specialist in dermatology at Stockholm's Diagnostic Centre for Skin Conditions. She has conducted research into dry skin, various skin conditions including atopic eczema and ichthyosis, and the effects of various emollient creams on dry skin. We first met when we were working together on a clinical study of the amino acid N-acetyl aspartic acid. I am grateful to Torborg for fact-checking this book. During the autumn of 2018, we lunched together frequently and had many inspiring conversations about the skin and how best to take care of it. Thank you, Torborg!

Marie Lodén

Marie is a pharmacist, lecturer, skin scientist, and skincare expert. She holds a doctorate on the impact of skincare products on the skin. She was head of research and development at ACO for many years, has written books about the skin, skincare and cosmetics, and now runs her own company, the Eviderm Institute, which provides consultancy services for development and studies of skincare products. Marie and I have met frequently over the last few years: sometimes on the courses she runs, to which I've been invited as a guest speaker, and sometimes in the context of development projects (in probiotics, for instance), or at dermatology conferences. We've had many meetings and discussions together, and although our views sometimes differ, Marie played a vital role in inspiring me to write my own book.

Lars Norlén

Lars is a dermatologist and lecturer at the Karolinska University Hospital's Skin Clinic and also runs a dermatology clinic at Sturebadet Health Care. We first met ten years ago at a dermatology conference in Copenhagen. Lars is a world expert in the skin barrier function. He and his team have used a highly sophisticated technology, cryo-electron microscopy, to investigate how the skin barrier is structured at atomic level. The better we understand the structure of the skin, the more we can learn about how to develop new molecules and new treatments. I am grateful to Lars for agreeing to be interviewed and for the discussion we had while I was working on this book.

Desmond Tobin

Professor Desmond Tobin of University College Dublin is a world expert in hair, the biology of hair, and alopecia. It would be hard to find a more deeply committed scientist. I first met him in the corridors of Bradford University, where he was in charge of a team researching the hair follicle and alopecia. He provided advice while I was working on *The Scandinavian Skincare Bible*. Desmond is now a professor at the Charles Institute of Dermatology, University College Dublin.

Noel Dew

Noel is a pharmacist with a doctorate in pharmacy and topical formulations for the skin from the University of Uppsala. He has been working in the field of skincare for a decade. I asked Noel to check the sources of the chapter on skincare — and he did so with aplomb. He was incredibly meticulous in examining every detail and used his expert knowledge to deal with the issues that arose. Thank you, Noel!

And finally …

I started writing *The Scandinavian Skincare Bible* at the beginning of 2018, in Stockholm. Although I was used to writing scientific articles, I soon learned that writing a work of popular science posed new challenges. I took a few false turns on my journey of discovery, but luckily some amazing friends, colleagues, and family members gave me a helping hand.

First of all, I'd like to thank my wonderful friend Karin Magnusson for all her wise advice and creative ideas, and for editing the text. Without you, Karin, this book wouldn't have existed.

Ebba Barrett Bandh and Claes Ericson at Bookmark have also earned my deepest gratitude for believing in the book and in me from the very first, and for doing their utmost to make my work easier. I'd also like to thank Ina Schuppe Koistinen, for your beautiful watercolours; Johan Barrett, for your fantastic design, Olof Bandh, for all your illustrations of the skin; and Emanuel Holm, for your patient and thorough editing. Every word was weighed on the finest of scales.

Marcus Jacobsson — without you, there'd be no book!

Stephanie Matti — you're a rock!

Alain Mavon, Susanne Fabre, Lene Visdal-Johnsen, and Ia Khmaladze, I've appreciated working with you on so many fascinating research projects over the years.

I also want to thank my fantastic friend Alexandra Petersson for all your devoted support and wise advice — and for my incredible godson, Matteo! Thank you to my Dad, Håkan! And thank you to my wonderful friends Åsa Göthlin, Malin Dahlström, Helen Halmell, Kristina Lagerstedt, and Sara Strid for giving me so much support on my journey.

Thank you to my little brother, Johan Gillbro, for cheering me on when I needed it most.

And thank you to my dear Mum, Agneta, for being there!

ACKNOWLEDGEMENTS

Bibliography

About the Skin

Aboobacker S, Karthikeyan K. Melasma and Hypothyroidism: Conflicting Coexistence? *J Clin Exp Dermatol Res*. 2017.

Ancans J, Tobin DJ, Hoogduijn MJ, Smit NP, Wakamatsu K, Thody AJ. Melanosomal pH controls rate of melanogenesis, eumelanin/phaeomelanin ratio and melanosome maturation in melanocytes and melanoma cells. *Exp Cell Res*. 2001.

Baldini E, Odorisio T, Sorrenti S, Catania A, Tartaglia F, Carbotta G, et al. Vitiligo and autoimmune thyroid disorders. *Frontiers in Endocrinology*. 2017.

Berardesca E, Farage M, Maibach H. Sensitive skin: An overview. *International Journal of Cosmetic Science*. 2013.

Bhate K, Williams HC. Epidemiology of acne vulgaris. *British Journal of Dermatology*. 2013.

Brenner M, Hearing VJ. The protective role of melanin against UV damage in human skin. *Photochemistry and Photobiology*. 2008.

Castelo-Branco C, Figueras F, Martínez De Osaba MJ, Vanrell JA. Facial wrinkling in postmenopausal women. Effects of smoking status and hormone replacement therapy. *Maturitas*. 1998.

Cheong WK. Gentle cleansing and moisturizing for patients with atopic dermatitis and sensitive skin. *American Journal of Clinical Dermatology*. 2009.

Davern J, O'Donnell AT. Stigma predicts health-related quality of life impairment, psychological distress, and somatic symptoms in acne sufferers. *PLoS One*. 2018.

Duncan KO, Leffell DJ. Preoperative assessment of the elderly patient. *Dermatologic Clinics*. 1997.

Egeberg A, Hansen PR, Gislason GH, Thyssen JP. Clustering of autoimmune diseases in patients with rosacea. *J Am Acad Dermatol*. 2016.

Egeberg A, Thyssen JP, Wu JJ, Skov L. Risk of first-time and recurrent depression in patients with psoriasis — a population-based cohort study. *Br J Dermatol*. 2018.

Ernster VL, Grady D, Miike R, Black D, Selby J, Kerlikowske K. Facial wrinkling in men and women, by smoking status. *Am J Public Health*. 1995.

Fajuyigbe D, Lwin SM, Diffey BL, Baker R, Tobin DJ, Sarkany RPE, et al. Melanin distribution in human epidermis affords localized protection against DNA photodamage and concurs with skin cancer incidence difference in extreme phototypes. *FASEB J.* 2018.

Farage MA, Miller KW, Maibach HI. Textbook of aging skin. 2010.

Feelisch M, Kolb-Bachofen V, Liu D, Lundberg JO, Revelo LP, Suschek CV, et al. Is sunlight good for our heart? *European Heart Journal.* 2010.

Flament F, Bazin R, Laquieze S, Rubert V, Simonpietri E, Piot B. Effect of the sun on visible clinical signs of aging in Caucasian skin. *Clin Cosmet Investig Dermatol.* 2013.

Foolad N, Shi VY, Prakash N, Kamangar F, Sivamani RK. The association of the sebum excretion rate with melasma, erythematotelangiectatic rosacea, and rhytides. *Dermatol Online J.* 2015.

Gether L, Overgaard LK, Egeberg A, Thyssen JP. Incidence and prevalence of rosacea: a systematic review and meta-analysis. *British Journal of Dermatology.* 2018.

Ghodsi SZ, Orawa H, Zouboulis CC. Prevalence, severity, and severity risk factors of acne in high school pupils: A community-based study. *J Invest Dermatol.* 2009.

Gonzalez-Hinojosa D, Jaime-Villalonga A, Aguilar-Montes G, and LL-O. Demodex and rosacea: Is there a relationship? *Indian J Ophthalmol.* 2018.

Grove GL. Physiologic changes in older skin. *Dermatol Clin.* 1986.

Harvell JD, Maibach HI. Percutaneous absorption and inflammation in aged skin: A review. *Journal of the American Academy of Dermatology.* 1994.

Heisig M, Reich A. Psychosocial aspects of rosacea with a focus on anxiety and depression. *Clinical, Cosmetic and Investigational Dermatology.* 2018.

Hornsby PJ. Biosynthesis of DHEAS by the Human Adrenal Cortex and Its Age-Related Decline. *Ann N Y Acad Sci.* 1995.

Hua C, Boussemart L, Mateus C, Routier E, Boutros C, Cazenave H, et al. Association of vitiligo with tumor response in patients with metastatic melanoma treated with pembrolizumab. *JAMA Dermatology.* 2016.

Ighodaro OM, Akinloye OA. First line defence antioxidants-superoxide dismutase (SOD), catalase (CAT) and glutathione peroxidase (GPX): Their fundamental role in the entire antioxidant defence grid. *Alexandria J Med.* 2017.

Jablonski NG, Chaplin G. Human skin pigmentation as an adaptation to UV radiation. *Proc Natl Acad Sci.* 2010.

Jackson SM, Williams ML, Feingold KR, Elias PM, Francisco S. Pathobiology of the stratum corneum. *West J Med.* 1993.

Kadunce DP, Burr R, Gress R, Kanner R, Lyon JL, Zone JJ. Cigarette smoking: Risk factor for premature facial wrinkling. *Ann Intern Med.* 1991.

Lacey N, Russell-Hallinan A, Zouboulis CC, Powell FC. Demodex mites modulate sebocyte immune reaction: possible role in the pathogenesis of rosacea. *Br J Dermatol.* 2018.

Lavker RM, Zheng PS, Dong G. Morphology of aged skin. *Clin Geriatr Med.* 1989.

Léger D Saint, François AM, Lévêque JL, Stoudemayer TJ, Grove GL, Kligman AM. Age-associated changes in stratum corneum lipids and their relation to dryness. *Dermatology.* 1988.

Liu D, Fernandez BO, Hamilton A, Lang NN, Gallagher JMC, Newby DE, et al. UVA irradiation of human skin vasodilates arterial vasculature and lowers blood pressure independently of nitric oxide synthase. *J Invest Dermatol.* 2014.

Lutfi RJ, Fridmanis M, Misiunas AL, Pafume O, Gonzalez EA, Villemur JA, et al. Association of melasma with thyroid autoimmunity and other thyroidal abnormalities and their relationship to the origin of the melasma. *J Clin Endocrinol Metab*. 1985.

Machková L, Švadlák D, Dolečková I. A comprehensive in vivo study of Caucasian facial skin parameters on 442 women. *Archives of Dermatological Research*. 2018.

Martini F, Nath J, Bartholomew E. Fundamentals of Anatomy & Physiology. *Journal of Chemical Information and Modeling*. 2009.

Matuoka K, Hasegawa N, Namba M, Smith GJ, Mitsui Y. A decrease in hyaluronic acid synthesis by aging human fibroblasts leading to heparan sulfate enrichment and growth reduction. Aging Clin Exp Res. 1989.

McCallion R, Po ALW. Dry and photo–aged skin and manifestations management. *Journal of Clinical Pharmacy and Therapeutics*. 1993.

Merinville E, Grennan GZ, Gillbro JM, Mathieu J, Mavon A. Influence of facial skin ageing characteristics on the perceived age in a Russian female population. *Int J Cosmet Sci*. 2015;37.

Mi HS, Rhie GE, Park CH, Kyu HK, Kwang HC, Hee CE, et al. Modulation of collagen metabolism by the topical application of dehydroepiandrosterone to human skin. *J Invest Dermatol*. 2005.

Niepomniszcze H, Huaier Amad R. Skin disorders and thyroid diseases. *Journal of Endocrinological Investigation*. 2001.

Nouveau S, Bastien P, Baldo F, de Lacharriere O. Effects of topical DHEA on aging skin: A pilot study. *Maturitas*. 2008.

Oh JH, Kim YK, Jung JY, Shin Jeun, Chung JH. Changes in glycosaminoglycans and related proteoglycans in intrinsically aged human skin in vivo. *Experimental Dermatology*. 2011.

Ortonne JP. Pigmentary changes of the ageing skin. *Br J Dermatol*. 1990.

Paller A, Jaworski JC, Simpson EL, Boguniewicz M, Russell JJ, Block JK, et al. Major Comorbidities of Atopic Dermatitis: Beyond Allergic Disorders. *Am J Clin Dermatol*. 2018.

Pierard-Franchimont C, Pierard GE, Kligman AM. Rhythm of Sebum Excretion during the Menstrual Cycle. *Dermatologica*. 1991.

Plikus MV, Guerrero-Juarez CF, Ito M, Li YR, Dedhia PH, Zheng Y, et al. Regeneration of fat cells from myofibroblasts during wound healing. *Science* (80-). 2017.

Plonka PM, Passeron T, Brenner M, Tobin DJ, Shibahara S, Thomas A, et al. What are melanocytes really doing all day long...? *Exp Dermatol*. 2009.

Pochi PE, Strauss JS. Sebaceous gland response in man to the administration of testosterone, delta-4-androstenedione, and dehydroisoandrosterone. *J Invest Dermatol*. 1969.

Ponsonby A-L, Lucas RM, Mei IAF. UVR, Vitamin D and Three Autoimmune Diseases — Multiple Sclerosis, Type 1 Diabetes, Rheumatoid Arthritis. *Photochem Photobiol*. 2005.

Quaglino P, Marenco F, Osella-Abate S, Cappello N, Ortoncelli M, Salomone B, et al. Vitiligo is an independent favourable prognostic factor in stage III and IV metastatic melanoma patients: Results from a single-institution hospital-based observational cohort study. *Ann Oncol*. 2010.

Rathi S, Achar A. Melasma: A clinico-epidemiological study of 312 cases. *Indian J Dermatol*. 2011.

Salem I, Ramser A, Isham N, Ghannoum MA. The gut microbiome as a major regulator of the gut-skin axis. *Frontiers in Microbiology*. 2018.

BIBLIOGRAPHY

Savoye I, Olsen CM, Whiteman DC, Bijon A, Wald L, Dartois L, et al. Patterns of ultraviolet radiation exposure and skin cancer risk: the E3N-SunExp Study. *J Epidemiol*. 2018.

Schallreuter KU. Advances in melanocyte basic science research. *Dermatol Clin*. 2007;25:283–291, vii.

Schallreuter KU, Moore J, Behrens-Williams S, Panske A, Harari M. Rapid initiation of repigmentation in vitiligo with Dead Sea climatotherapy in combination with pseudocatalase (PC-KUS). *International Journal of Dermatology*. 2002. pp. 482–7.

Schallreuter KU, Moore J, Wood JM, Beazley WD, Gaze DC, Tobin DJ, et al. In vivo and in vitro evidence for hydrogen peroxide (H2O2) accumulation in the epidermis of patients with vitiligo and its successful removal by a UVB-activated pseudocatalase. *J Investig Dermatol Symp Proc*. 1999;4:91–6.

Schallreuter KU, Moore J, Wood JM, Beazley WD, Peters EM, Marles LK, et al. Epidermal H(2)O(2) accumulation alters tetrahydrobiopterin (6BH4) recycling in vitiligo: identification of a general mechanism in regulation of all 6BH4-dependent processes? *J Invest Dermatol*. 2001;116:167–74.

Schallreuter KU, Rokos H, Chavan B, Gillbro JM, Cemeli E, Zothner C, et al. Quinones are reduced by 6-tetrahydrobiopterin in human keratinocytes, melanocytes, and melanoma cells. *Free Radic Biol Med*. 2008;44(4).

Schallreuter KU, Wood JM, Lemke KR, Levenig C. Treatment of vitiligo with a topical application of pseudocatalase and calcium in combination with short-term UVB exposure: a case study on 33 patients. *Dermatology*. 1995;190:223–9.

Schwarz ST, Xing Y, Tomar P, Bajaj N, Auer DP. In Vivo Assessment of Brainstem Depigmentation in Parkinson Disease: Potential as a Severity Marker for Multicenter Studies. *Radiology*. 2017.

Simonsick EM, Meier HCS, Shaffer NC, Studenski SA, Ferrucci L. Basal body temperature as a biomarker of healthy aging. *Age* (Omaha). 2016.

Slominski AT, Zmijewski MA, Plonka PM, Szaflarski JP, Paus R. How UV Light Touches the Brain and Endocrine System Through Skin, and Why. *Endocrinology*. 2018.

Tobin DJ. Introduction to skin aging. *J Tissue Viability*. 2017.

Tobin DJ, Swanson NN, Pittelkow MR, Peters EM, Schallreuter KU. Melanocytes are not absent in lesional skin of long duration vitiligo. *J Pathol*. 2000;191:407–16.

Tur E, Oren-Vulfs S, Yosipovitch G. Chronic and Acute Effects of Cigarette Smoking on Skin Blood Flow. *Angiology*. 1992.

Waller JM, Maibach HI. Age and skin structure and function, a quantitative approach (I): Blood flow, pH, thickness, and ultrasound echogenicity. *Skin Research and Technology*. 2005.

Wang AS, Dreesen O. Biomarkers of cellular senescence and skin aging. *Frontiers in Genetics*. 2018.

Weller RB. Sunlight Has Cardiovascular Benefits Independently of Vitamin D. *Blood Purification*. 2016.

Whiteman DC, Parsons PG, Green AC. Determinants of melanocyte density in adult human skin. *Arch Dermatol Res*. 1999.

Wright F, Weller RB. Risks and benefits of UV radiation in older people: More of a friend than a foe? *Maturitas*. 2015.

BIBLIOGRAPHY

About Skincare

Ackermann L, Virtanen H, Korhonen L, Laukkanen A, Huilaja L, Riekki R, et al. An epidemic of allergic contact dermatitis caused by a new allergen, caprylhydroxamic acid, in moisturizers. *Contact Dermatitis*. 2017.

Al-Bader T, Byrne A, Gillbro J, Mitarotonda A, Metois A, Vial F, et al. Effect of cosmetic ingredients as anticellulite agents: Synergistic action of actives with in vitro and in vivo efficacy. *J Cosmet Dermatol*. 2012.

Al-Niaimi F, Chiang NYZ. Topical Vitamin C and the Skin: Mechanisms of Action and Clinical Applications. *J Clin Aesthet Dermatol*. 2017.

Andersen F, Hedegaard K, Petersen TK, Bindslev-Jensen C, Fullerton A, Andersen KE. Comparison of the effect of glycerol and triamcinolone acetonide on cumulative skin irritation in a randomized trial. *J Am Acad Dermatol*. 2007.

Arif T. Salicylic acid as a peeling agent: A comprehensive review. *Clinical, Cosmetic and Investigational Dermatology*. 2015.

Bernstein EF, Brown DB, Schwartz MD, Kaidbey K, Ksenzenko SM, Draelos ZD. The Polyhydroxy Acid Gluconolactone Protects Against Ultraviolet Radiation in an In Vitro Model of Cutaneous Photoaging. *Dermatologic Surg*. 2004.

Binder S, Hanáková A, Tománková K, Pízová K, Bajgar R, Manišová B, et al. Adverse phototoxic effect of essential plant oils on nih 3T3 cell line after UV light exposure. *Cent Eur J Public Health*. 2016.

Bosch R, Philips N, Suárez-Pérez J, Juarranz A, Devmurari A, Chalensouk-Khaosaat J, et al. Mechanisms of Photoaging and Cutaneous Photocarcinogenesis, and Photoprotective Strategies with Phytochemicals. *Antioxidants*. 2015.

Bylka W, Znajdek-Awizeń P, Studzińska-Sroka E, Dańczak-Pazdrowska A, Brzezińska M. Centella asiatica in dermatology: An overview. *Phytotherapy Research*. 2014.

Campos D, Gravato C, Fedorova G, Burkina V, Soares AMVM, Pestana JLT. Ecotoxicity of two organic UV-filters to the freshwater caddisfly Sericostoma vittatum. *Environ Pollut*. 2017.

Chlebus E, Chlebus M. Factors affecting the course and severity of adult acne. Observational cohort study. *J Dermatolog Treat*. 2017.

Conti A, Rogers J, Verdejo P, Harding CR, Rawlings AV. Seasonal influences on stratum corneum ceramide 1 fatty acids and the influence of topical essential fatty acids. *Int J Cosmet Sci*. 1996.

Denda S, Denda M, Inoue K, Hibino T. Glycolic acid induces keratinocyte proliferation in a skin equivalent model via TRPV1 activation. *J Dermatol Sci*. 2010.

Dhaliwal S, Rybak I, Ellis SR, Notay M, Trivedi M, Burney W, Vaughn AR, Nguyen M, Reiter P, Bosanac S, Yan H, Foolad N, Sivamani RK. Prospective, randomized, double-blind assessment of topical bakuchiol and retinol for facial photoaging. *Br J Dermatology*. 2018.

Di Nardo A, Wertz P, Giannetti A, Seidenari S. Ceramide and cholesterol composition of the skin of patients with atopic dermatitis. *Acta Derm Venereol*. 1998.

Didierjean L, Carraux P, Grand D, Sass JO, Nau H, Saurat JH. Topical retinaldehyde increases skin content of retinoic acid and exerts biologic activity in mouse skin. *J Invest Dermatol*. 1996.

Dijoux N, Guingand Y, Bourgeois C, Durand S, Fromageot C, Combe C, et al. Assessment of the phototoxic hazard of some essential oils using modified 3T3 neutral red uptake assay. *Toxicol Vitr*. 2006.

Draelos ZD, Matsubara A, Smiles K. The effect of 2% niacinamide on facial sebum production. *J Cosmet Laser Ther*. 2006.

Feelisch M, Kolb-Bachofen V, Liu D, Lundberg JO, Revelo LP, Suschek CV, et al. Is sunlight good for our heart? *European Heart Journal*. 2010.

Fitzpatrick RE, Rostan EF. Double-blind, half-face study comparing topical vitamin C and vehicle for rejuvenation of photodamage. *Dermatologic Surg*. 2002.

Fu PP, Xia QS, Yin JJ, Cherng S-HH, Yan J, Mei N, et al. Photodecomposition of Vitamin A and Photobiological Implications for the Skin. *Photochem Photobiol*. 2007.

Fulton JE. Comedogenicity and irritancy of commonly used ingredients in skin care products. *J Soc Cosmet Chem*. 1989.

Gaspar LR, Tharmann J, Maia Campos PMBG, Liebsch M. Skin phototoxicity of cosmetic formulations containing photounstable and photostable UV-filters and vitamin A palmitate. *Toxicol Vitr*. 2013.

Giacomoni PU, Declercq L, Hellemans L, Maes D. Aging of human skin: review of a mechanistic model and first experimental data. *IUBMB Life*. 2000.

Gillbro JM, Olsson MJ. The melanogenesis and mechanisms of skin-lightening agents-existing and new approaches. *Int J Cosmet Sci* [Internet]. 2011;33:210–21. Available from: http://www.ncbi.nlm.nih.gov/pubmed/21265866.

Goodman H. *Cosmetic Dermatology*. 1st ed. McGraw-Hill. New York; 1936.

Gressel KL, Duncan FJ, Oberyszyn TM, La Perle KM, Everts HB. Endogenous Retinoic Acid Required to Maintain the Epidermis Following Ultraviolet Light Exposure in SKH-1 Hairless Mice. *Photochem Photobiol*. 2015.

Harvey PW, Darbre P. Endocrine disrupters and human health: Could oestrogenic chemicals in body care cosmetics adversely affect breast cancer incidence in women? A review of evidence and call for further research. *Journal of Applied Toxicology*. 2004.

Heuschkel S, Wohlrab J, Neubert RHH. Dermal and transdermal targeting of dihydroavenanthramide D using enhancer molecules and novel microemulsions. *Eur J Pharm Biopharm*. 2009.

Hu S, Belcaro G, Dugall M, Hosoi M, Togni S, Maramaldi G, et al. Aescin-based topical formulation to prevent foot wounds and ulcerations in diabetic microangiopathy. *Eur Rev Med Pharmacol Sci*. 2016.

Huth S, Schmitt L, Marquardt Y, Heise R, Lüscher B, Amann PM, et al. Effects of a ceramide-containing water-in-oil ointment on skin barrier function and allergen penetration in an IL-31 treated 3D model of the disrupted skin barrier. *Experimental Dermatology*. 2018.

Jacob SE, Scheman A, McGowan MA. Propylene Glycol. *Dermatitis*. 2018.

Khanna S, Dash PR, Darbre PD. Exposure to parabens at the concentration of maximal proliferative response increases migratory and invasive activity of human breast cancer cells in vitro. *J Appl Toxicol*. 2014.

Kong R, Cui Y, Fisher GJ, Wang X, Chen Y, Schneider LM, et al. A comparative study of the effects of retinol and retinoic acid on histological, molecular, and clinical properties of human skin. *Journal of Cosmetic Dermatology*. 2015.

Krause M, Klit A, Blomberg Jensen M, Søeborg T, Frederiksen H, Schlumpf M, et al. Sunscreens: Are they beneficial for health? An overview of endocrine disrupting properties of UV-filters. *International Journal of Andrology*. 2012.

Krüger C, Schallreuter KU. Stigmatisation, avoidance behaviour and difficulties in coping are common among adult patients with vitiligo. *Acta Derm Venereol.* 2015.

Lalla SC, Nguyen H, Chaudhry H, Killian JM, Drage LA, Davis MDP, Yiannias JA, Hall MR. Patch Testing to Propylene Glycol: The Mayo Clinic Experience. *Dermatitis.* 2018;Jul/Aug;29:200–5.

Lee SH, Huh CH, Park KC, Youn SW. Effects of repetitive superficial chemical peels on facial sebum secretion in acne patients. *J Eur Acad Dermatology Venereol.* 2006.

Li DG, Du HY, Gerhard S, Imke M, Liu W. Inhibition of TRPV1 prevented skin irritancy induced by phenoxyethanol. A preliminary in vitro and in vivo study. *Int J Cosmet Sci.* 2017.

Lio PA. Little white spots: An approach to hypopigmented macules. *Archives of Disease in Childhood: Education and Practice Edition.* 2008.

Liu D, Fernandez BO, Hamilton A, Lang NN, Gallagher JMC, Newby DE, et al. UVA irradiation of human skin vasodilates arterial vasculature and lowers blood pressure independently of nitric oxide synthase. *J Invest Dermatol.* 2014.

Liu H, Yu H, Xia J, Liu L, Liu GJ, Sang H. Topical azelaic acid, salicylic acid, nicotinamide, and sulphur for acne. *Cochrane Database Syst Rev.* 2014.

Lodén M. Role of Topical Emollients and Moisturizers in the Treatment of Dry Skin Barrier Disorders. *American Journal of Clinical Dermatology.* 2003.

Lodén M, Andersson AC, Anderson C, Bergbrant IM, Frödin T, Öhman H, et al. A double-blind study comparing the effect of glycerin and urea on dry, eczematous skin in atopic patients. *Acta Derm Venereol.* 2002.

Luzzi R, Feragalli B, Belcaro G, Cesarone MR, Cornelli U, Dugall M, et al. Aescin: Microcirculatory activity. Effects of accessory components on clinical and microcirculatory efficacy. *Panminerva Med.* 2011.

Lykkesfeldt G, Hoyer H. Topical Choloesterol Treatment of Recessive X-linked Ichthyosis. *Lancet.* 1983.

McDaniel DH, Mazur C, Wortzman MS, Nelson DB. Efficacy and tolerability of a double-conjugated retinoid cream vs 1.0% retinol cream or 0.025% tretinoin cream in subjects with mild to severe photoaging. *J Cosmet Dermatol.* 2017.

McGowan MA, Scheman A, Jacob SE. Propylene Glycol in Contact Dermatitis: A Systematic Review. *Dermatitis.* 2018.

Merinville E, Grennan GZ, Gillbro JM, Mathieu J, Mavon A. Influence of facial skin ageing characteristics on the perceived age in a Russian female population. *Int J Cosmet Sci.* 2015.

Milani M, Sparavigna A. The 24-hour skin hydration and barrier function effects of a hyaluronic 1%, glycerin 5%, and Centella asiatica stem cells extract moisturizing fluid: An intra-subject, randomized, assessor-blinded study. *Clin Cosmet Investig Dermatol.* 2017.

Mizukawa A, Molins-Delgado D, de Azevedo JCR, Fernandes CVS, Díaz-Cruz S, Barceló D. Sediments as a sink for UV filters and benzotriazoles: the case study of Upper Iguaçu watershed, Curitiba (Brazil). *Environ Sci Pollut Res.* 2017.

Nasrollahi SA, Ayatollahi A, Yazdanparast T, Samadi A, Hosseini H, Shamsipour M, et al. Comparison of linoleic acid-containing water-in-oil emulsion with urea-containing water-in-oil emulsion in the treatment of atopic dermatitis: a randomized clinical trial. *Clin Cosmet Investig Dermatol.* 2018.

Newman N, Newman A, Moy LS, Babapour R, Harris AG, Moy RL. Clinical improvement of photoaged skin with 50% glycolic acid: A double-blind vehicle-controlled study. *Dermatologic Surg.* 1996.

Nguyen SH, Dang TP, Maibach HI. Comedogenicity in rabbit: Some cosmetic ingredients/vehicles. *Cutan Ocul Toxicol.* 2007.

Norlén L. Is Oil a Balsam for Baby Skin? *Acta dermato-venereologica.* 2016.

Oh M, Lee J, Kim S, Cho SA, Lee E, Yeon JH, et al. Regional and seasonal differences in skin irritation and neurosensitivity in Chinese and South Korean women. *J Eur Acad Dermatology Venereol.* 2015.

Panin G, Strumia R, Ursini F. Topical α-tocopherol acetate in the bulk phase: Eight years of experience in skin treatment. *Annals of the New York Academy of Sciences.* 2004.

Pathan IB, Setty M. Chemical Penetration Enhancers for Transdermal Drug Delivery Systems. *Trop J Pharm Res.* 2009.

Piazena H, Pittermann W, Müller W, Jung K, Kelleher DK, Herrling T, et al. Effects of water-filtered infrared-A and of heat on cell death, inflammation, antioxidative potential and of free radical formation in viable skin — First results. *J Photochem Photobiol B Biol.* 2014.

Ponsonby A-L, Lucas RM, Mei IAF. UVR, Vitamin D and Three Autoimmune Diseases — Multiple Sclerosis, Type 1 Diabetes, Rheumatoid Arthritis. *Photochem Photobiol.* 2005.

Puviani M, Tovecci F, Milani M. A two-center, assessor-blinded, prospective trial evaluating the efficacy of a novel hypertonic draining cream for cellulite reduction: A Clinical and instrumental (Antera 3D CS) assessment. *J Cosmet Dermatol.* 2018.

Quay ER, Chang YC, Graber E. Evidence for Anti-Aging South Korean Cosmeceuticals. *J Drugs Dermatol.* 2017.

Re DE, Whitehead RD, Xiao D, Perrett DI. Oxygenated-blood colour change thresholds for perceived facial redness, health, and attractiveness. *PLoS One.* 2011.

Rocha MA, Bagatin E. Skin barrier and microbiome in acne. *Archives of Dermatological Research.* 2018.

Schallreuter KU. Q10-triggered facial vitiligo. *Br J Dermatol.* 2013.

Schallreuter KU, Wood JM, Pittelkow MR, Gütlich M, Lemke KR, Rödl W, et al. Regulation of melanin biosynthesis in the human epidermis by tetrahydrobiopterin. *Science* (80–). 1994.

Serrano G, Almudéver P, Serrano JM, Milara J, Torrens A, Expósito I, et al. Phosphatidylcholine liposomes as carriers to improve topical ascorbic acid treatment of skin disorders. *Clin Cosmet Investig Dermatol.* 2015.

Shao Y, He T, Fisher GJ, Voorhees JJ, Quan T. Molecular basis of retinol anti-ageing properties in naturally aged human skin in vivo. *Int J Cosmet Sci.* 2017.

Slominski AT, Zmijewski MA, Plonka PM, Szaflarski JP, Paus R. How UV Light Touches the Brain and Endocrine System Through Skin, and Why. *Endocrinology.* 2018.

Stringer T, Nagler A, Orlow SJ, Oza VS. Clinical evidence for washing and cleansers in acne vulgaris: a systematic review. *Journal of Dermatological Treatment.* 2018.

Szél E, Polyánka H, Szabó K, Hartmann P, Degovics D, Balázs B, et al. Anti-irritant and anti-inflammatory effects of glycerol and xylitol in sodium lauryl sulphate-induced acute irritation. *J Eur Acad Dermatology Venereol.* 2015.

Tobin DJ, Swanson NN, Pittelkow MR, Peters EM, Schallreuter KU. Melanocytes are not absent in lesional skin of long duration vitiligo. *J Pathol*. 2000;191:407–16.

Tolleson WH, Cherng SH, Xia Q, Boudreau M, Yin JJ, Wamer WG, et al. Photodecomposition and phototoxicity of natural retinoids. *International Journal of Environmental Research and Public Health*. 2005.

Uter W, Werfel T, White IR, Johansen JD. Contact Allergy: A Review of Current Problems from a Clinical Perspective. *Int J Environ Res Public Health*. 2018.

Verheyen GR, Ooms T, Vogels L, Vreysen S, Bovy A, Van Miert S, Meersman F. Insects as an Alternative Source for the Production of Fats for Cosmetics. *J Cosmet Sci*. May/Jun; 69(3)187–202. 2018.

Wang J, Pan L, Wu S, Lu L, Xu Y, Zhu Y, et al. Recent advances on endocrine disrupting effects of UV filters. *International Journal of Environmental Research and Public Health*. 2016.

Weller RB. Sunlight Has Cardiovascular Benefits Independently of Vitamin D. *Blood Purification*. 2016.

Wright F, Weller RB. Risks and benefits of UV radiation in older people: More of a friend than a foe? *Maturitas*. 2015.

Yanase K, Hatta I. Disruption of human stratum corneum lipid structure by sodium dodecyl sulphate. *Int J Cosmet Sci*. 2017.

About the Microbiome

Abrahamsson TR, Jakobsson HE, Andersson AF, Björkstén B, Engstrand L, Jenmalm MC. Low gut microbiota diversity in early infancy precedes asthma at school age. *Clin Exp Allergy*. 2014.

Allhorn M, Arve S, Brüggemann H, Lood R. A novel enzyme with antioxidant capacity produced by the ubiquitous skin colonizer Propionibacterium acnes. *Sci Rep*. 2016.

Belkacemi S, Cassir N, Delerce J, Cadoret F, La Scola B. Bacteroides cutis, a new bacterial species isolated from human skin. *New Microbes New Infect*. 2018.

Bjerre RD, Bandier J, Skov L, Engstrand L, Johansen JD. The role of the skin microbiome in atopic dermatitis: a systematic review. *British Journal of Dermatology*. 2017.

Bowe WP, Logan AC, Uhlenhake E, Yentzer B, Feldman S, Loney T, et al. Acne vulgaris, probiotics and the gut-brain-skin axis — back to the future? *Gut Pathog*. 2011.

Capone KA, Dowd SE, Stamatas GN, Nikolovski J. Diversity of the human skin microbiome early in life. *J Invest Dermatol*. 2011.

Celiker H. A new proposed mechanism of action for gastric bypass surgery: Air hypothesis. *Med Hypotheses*. 2017.

Chassaing B, Koren O, Goodrich JK, Poole AC, Srinivasan S, Ley RE, et al. Dietary emulsifiers impact the mouse gut microbiota promoting colitis and metabolic syndrome. *Nature*. 2015.

Chen HW, Liu PF, Liu YT, Kuo S, Zhang XQ, Schooley RT, et al. Nasal commensal Staphylococcus epidermidis counteracts influenza virus. *Sci Rep*. 2016.

Cogen AL, Nizet V, Gallo RL. Skin microbiota: A source of disease or defence? *Br J Dermatol*. 2008.

Cosseau C, Romano-Bertrand S, Duplan H, Lucas O, Ingrassia I, Pigasse C, et al. Proteobacteria from the human skin microbiota: Species-level diversity and hypotheses. *One Heal*. 2016.

Deidda F, Amoruso A, Nicola S, Graziano T, Pane M, Mogna L. New Approach in Acne Therapy: A Specific Bacteriocin Activity and a Targeted Anti IL-8 Property in Just 1 Probiotic Strain, the L. salivarius LS03. *J Clin Gastroenterol*. 2018.

Dermatology AA of. http://www.aad.org/stories-and-news/news-releases/could-probiotics-be-the-next-big-thing-in-acne-and-rosacea-treatments. Schaumburg. 2014.

Dreno B, Martin R, Moyal D, Henley JB, Khammari A, Seité S. Skin microbiome and acne vulgaris: Staphylococcus, a new actor in acne. *Exp Dermatol*. 2017.

Fabbrocini G, Bertona M, Picazo, Pareja-Galeano H, Monfrecola G, Emanuele E. Supplementation with Lactobacillus rhamnosus SP1 normalises skin expression of genes implicated in insulin signalling and improves adult acne. *Benef Microbes*. 2016.

Fahlén A, Engstrand L, Baker BS, Powles A, Fry L. Comparison of bacterial microbiota in skin biopsies from normal and psoriatic skin. *Arch Dermatol Res*. 2012.

Flohr C, Yeo L. Atopic dermatitis and the hygiene hypothesis revisited. *Curr Probl Dermatol*. 2011.

Fyhrquist N, Ruokolainen L, Suomalainen A, Lehtimäki S, Veckman V, Vendelin J, et al. Acinetobacter species in the skin microbiota protect against allergic sensitization and inflammation. *J Allergy Clin Immunol*. 2014.

Ganju G. Microbial community profiling shows dysbiosis in the lesional skin of Vitiligo subjects. *Sci rep*. 2016;13(6).

Grice EA, Kong HH, Conlan S, Deming CB, Davis J, Young AC, et al. Topographical and temporal diversity of the human skin microbiome. *Science*. 2009.

Grönroos M, Parajuli A, Laitinen OH, Roslund MI, Vari HK, Hyöty H, et al. Short-term direct contact with soil and plant materials leads to an immediate increase in diversity of skin microbiota. *MicrobiologyOpen*. 2018.

Gueniche A, Knaudt B, Schuck E, Volz T, Bastien P, Martin R, et al. Effects of nonpathogenic gram-negative bacterium Vitreoscilla filiformis lysate on atopic dermatitis: A prospective, randomized, double-blind, placebo-controlled clinical study. *Br J Dermatol*. 2008.

Gueniche A, Philippe D, Bastien P, Buyukpamukcu E, Reygagne P, Castiel I. Oral supplementation with probiotic Lactobacillus paracasei ST-11 improves dandruff condition. *Int J Trichology*. 2011.

Hacini-Rachinel F, Gheit H, Le Luduec JB, Dif F, Nancey S, Kaiserlian D. Oral probiotic control skin inflammation by acting on both effector and regulatory T cells. *PLoS One*. 2009.

Han XD, Oon HH, Goh CL. Epidemiology of post-adolescence acne and adolescence acne in Singapore: a 10-year retrospective and comparative study. *J Eur Acad Dermatology Venereol*. 2016.

Hong YH, Chang UJ, Kim YS, Jung EY, Suh HJ. Dietary galacto-oligosaccharides improve skin health: A randomized double blind clinical trial. *Asia Pac J Clin Nutr*. 2017.

Hospodsky D, Pickering AJ, Julian TR, Miller D, Gorthala S, Boehm AB, et al. Hand bacterial communities vary across two different human populations. *Microbiol* (United Kingdom). 2014.

Irwin SV, Fisher P, Graham E, Malek A, Robidoux A. Sulfites inhibit the growth of four species of beneficial gut bacteria at concentrations regarded as safe for food. *PLoS One*. 2017.

Jesenak M, Urbancek S, Majtan J, Banovcin P, Hercogova J. β-Glucan-based cream (containing pleuran isolated from pleurotus ostreatus) in supportive treatment of mild-to-moderate atopic dermatitis. *J Dermatolog Treat*. 2016.

Jo JH, Deming C, Kennedy EA, Conlan S, Polley EC, Ng W lan, et al. Diverse Human Skin Fungal Communities in Children Converge in Adulthood. *J Invest Dermatol*. 2016.

Jugé R, Rouaud-Tinguely P, Breugnot J, Servaes K, Grimaldi C, Roth M-P, et al. Shift in skin microbiota of Western European women across aging. *J Appl Microbiol* [Internet]. 2018;0–3. Available from: http://doi.wiley.com/10.1111/jam.13929.

Kang BS, Seo JG, Lee GS, Kim JH, Kim SY, Han YW, et al. Antimicrobial activity of enter-ocins from Enterococcus faecalis SL-5 against Propionibacterium acnes, the causative agent in acne vulgaris, and its therapeutic effect. *J Microbiol*. 2009.

Kaur S, Kaur S. Bacteriocins as potential anti-cancer agents. *Frontiers in Pharmacology*. 2015.

Kim HN, Yun Y, Ryu S, Chang Y, Kwon MJ, Cho J, et al. Correlation between gut microbiota and personality in adults: A cross-sectional study. *Brain Behav Immun*. 2018.

Kim J, Lee BS, Kim B, Na I, Lee J, Lee JY, et al. Identification of atopic dermatitis phenotypes with good responses to probio-tics (Lactobacillus plantarum CJLP133) in children. *Benef Microbes*. 2017.

Kong HH, Segre JA. Skin microbiome: looking back to move forward. *J Invest Dermatol* [Internet]. 2012;132(3 Pt 2):933–9. Available from: http://www.pubmedcentral.nih.gov/articlerender.fcgi?artid=3279608&tool=pmcentrez&ren-dertype=abstract.

Lee DE, Huh C-S, Ra J, Choi I-D, Jeong J-W, Kim S-H, et al. Clinical Evidence of Effects of Lactobacillus plantarum HY7714 on Skin Aging: A Randomized, Double Blind, Placebo-Controlled Study. *J Microbiol Biotechnol*. 2015.

Maarouf M, Platto JF, Shi VY. The role of nutrition in inflammatory pilosebaceous disorders: Implication of the skin-gut axis. *Australas J Dermatol*. 2018.

Mammone T, Gan D, Fthenakis C, Marenus K. The effect of N-acetyl-glucosamine on stratum corneum desquamation and water content in human skin. *Int J Cosmet Sci*. 2010.

Mori N, Kano M, Masuoka N, Konno T, Suzuki Y, Miyazaki K, et al. Effect of probiotic and prebiotic fermented milk on skin and intestinal conditions in healthy young female students. *Biosci Microbiota, Food Heal*. 2016.

Mottin VHM, Suyenaga ES. An approach on the potential use of probiotics in the treatment of skin conditions: acne and atopic dermatitis. *Int J Dermatol*. 2018.

Muizzuddin N, Maher W, Sullivan M, Schnittger S, Mammone T. Physiological effect of a probiotic on skin. *J Cosmet Sci*. 2012.

Myles IA, Uzel G, Datta SK. First-in-human topical microbiome transplantation with Roseomonas mucosa for atopic dermatitis. *JCI Insight*. 2018.

Nakatsuji T, Chen TH, Butcher AM, Trzoss LL, Nam SJ, Shirakawa KT, et al. A com-mensal strain of Staphylococcus epidermidis protects against skin neoplasia. *Sci Adv*. 2018.

Nam JH, Yun Y, Kim HS, Kim HN, Jung HJ, Chang Y, et al. Rosacea and its association with enteral microbiota in Korean females. *Exp Dermatol*. 2018.

Nutten S. Atopic dermatitis: Global epidemiology and risk factors. *Ann Nutr Metab*. 2015.

O'Neill AM, Gallo R. Host-microbiome interactions and recent progress into understanding the biology of acne vulgaris. *Microbiome*. 2018.

O'Neill CA, Monteleone G, McLaughlin JT, Paus R. The gut-skin axis in health and disease: A paradigm with therapeutic implications. *BioEssays*. 2016.

Oh J, Conlan S, Polley EC, Segre JA, Kong HH. Shifts in human skin and nares microbiota of healthy children and adults. *Genome Med*. 2012.

Okada H, Kuhn C, Feillet H, Bach J-F. The hygiene hypothesis for autoimmune and allergic diseases: an update. *Clin Exp Immunol*. 2010.

Pieterse R, Todorov SD. Bacteriocins: Exploring alternatives to antibiotics in mastitis treatment. *Brazilian Journal of Microbiology*. 2010.

Rao SSC, Rehman A, Yu S, De Andino NM. Brain fogginess, gas and bloating: A link between SIBO, probiotics and metabolic acidosis article. *Clinical and Translational Gastroenterology*. 2018.

Reygagne P, Bastien P, Couavoux MP, Philippe D, Renouf M, Castiel-Higounenc I, et al. The positive benefit of Lactobacillus paracasei NCC2461 ST11 in healthy volunteers with moderate to severe dandruff. *Benef Microbes*. 2017.

Rocha MA, Bagatin E. Skin barrier and microbiome in acne. *Archives of Dermatological Research*. 2018.

Ross AA, Doxey AC, Neufeld JD. The Skin Microbiome of Cohabiting Couples. *mSystems*. 2017.

Sanford JA, Gallo RL. Functions of the skin microbiota in health and disease. *Seminars in Immunology*. 2013.

Saunders CW, Scheynius A, Heitman J. Malassezia fungi are specialized to live on skin and associated with dandruff, eczema, and other skin diseases. *PLoS Pathog*. 2012.

Seité S, Zelenkova H, Martin R. Clinical efficacy of emollients in atopic dermatitis patients — relationship with the skin microbiota modification. *Clin Cosmet Investig Dermatol*. 2017.

Shibagaki N, Suda W, Clavaud C, Bastien P, Takayasu L, Iioka E, et al. Aging-related changes in the diversity of women's skin microbiomes associated with oral bacteria. *Sci Rep*. 2017.

Sikorska H, Smoragiewicz W. Role of probiotics in the prevention and treatment of meticillin-resistant Staphylococcus aureus infections. *International Journal of Antimicrobial Agents*. 2013.

Song H, Yoo Y, Hwang J, Na YC, Kim HS. Faecalibacterium prausnitzii subspecies-level dysbiosis in the human gut microbiome underlying atopic dermatitis. *J Allergy Clin Immunol*. 2016.

Song SJ, Lauber C, Costello EK, Lozupone CA, Humphrey G, Berg-Lyons D, et al. Cohabiting family members share microbiota with one another and with their dogs. *Elife*. 2013.

Sparber F, LeibundGut-Landmann S. Host responses to Malassezia spp. in the mammalian skin. *Frontiers in Immunology*. 2017.

Stuart E. Amway and Microbiome Insights Identify Bacteria That Could Improve Skin Care. *Cosmet Toilet*. 2017.

BIBLIOGRAPHY

Sun H, Chen Y, Cheng M, Zhang X, Zheng X, Zhang Z. The modulatory effect of polyphenols from green tea, oolong tea and black tea on human intestinal microbiota in vitro. *J Food Sci Technol*. 2018.

Teo WY, Chan MY, Lam CM, Chong CY. Skin manifestation of Stenotrophomonas maltophilia infection — a case report and review article. *Ann Acad Med Singapore*. 2006.

Wang Q, Cui S, Zhou L, He K, Song L, Liang H, He C. Effect of cosmetic chemical preservatives on resident flora isolated from healthy facial skin. *J Cosmet Dermatol*. 2018.

Wang Y, Kuo S, Shu M, Yu J, Huang S, Dai A, et al. Staphylococcus epidermidis in the human skin microbiome mediates fermentation to inhibit the growth of Propionibacterium acnes: Implications of probiotics in acne vulgaris. *Appl Microbiol Biotechnol*. 2014;98(1): 411–24.

Watanabe S, Narisawa Y, Arase S, Okamatsu H, Ikenaga T, Tajiri Y, et al. Differences in fecal microflora between patients with atopic dermatitis and healthy control subjects. *J Allergy Clin Immunol*. 2003.

Weiss E, Katta R. Diet and rosacea: the role of dietary change in the management of rosacea. *Dermatol Pract Concept*. 2017.

Yatsunenko T, Rey FE, Manary MJ, Trehan I, Dominguez-Bello MG, Contreras M, et al. Human gut microbiome viewed across age and geography. *Nature*. 2012.

Ying S, Zeng D-N, Chi L, Tan Y, Galzote C, Cardona C, et al. The Influence of Age and Gender on Skin-Associated Microbial Communities in Urban and Rural Human Populations. *PLoS One*. 2015.

About Lifestyle

Abolaji AO, Olaiya CO, Oluwadahunsi OJ, Farombi EO. Dietary consumption of monosodium L-glutamate induces adaptive response and reduction in the life span of Drosophila melanogaster. *Cell Biochem Funct*. 2017.

Afifi L, Danesh MJ, Lee KM, Beroukhim K, Farahnik B, Ahn RS, et al. Dietary Behaviors in Psoriasis: Patient-Reported Outcomes from a U.S. National Survey. *Dermatol Ther* (Heidelb). 2017.

Agak GW, Qin M, Nobe J, Kim MH, Krutzik SR, Tristan GR, et al. Propionibacterium acnes induces an IL-17 response in acne vulgaris that is regulated by vitamin A and vitamin D. *J Invest Dermatol*. 2014.

Akdeniz M, Tomova-Simitchieva T, Dobos G, Blume-Peytavi U, Kottner J. Does dietary fluid intake affect skin hydration in healthy humans? A systematic literature review. *Skin Research and Technology*. 2018.

Amer M, Bahgat MR, Tosson Z, Abdel Mowla MY, Amer K. Serum zinc in acne vulgaris. *Int J Dermatol*. 1982.

Amon U, Baier L, Yaguboglu R, Ennis M, Holick MF, Amon J. Serum 25-hydroxyvitamin D levels in patients with skin diseases including psoriasis, infections, and atopic dermatitis. *Dermatoendocrinol*. 2018.

Asarch A, Barak O, Loo DS, Gottlieb AB. Th17 cells: a new therapeutic target in inflammatory dermatoses. *J Dermatolog Treat*. 2008.

Bhargava R, Kumar P, Kumar M, Mehra N, Mishra A. A randomized controlled trial of omega-3 fatty acids in dry eye syndrome. *Int J Ophthalmol*. 2013.

Bhate K, Williams HC. Epidemiology of acne vulgaris. *British Journal of Dermatology*. 2013.

Bhatia BK, Millsop JW, Debbaneh M, Koo J, Linos E, Liao W. Diet and psoriasis, part II: Celiac disease and role of a gluten-free diet. *Journal of the American Academy of Dermatology*. 2014.

Brescoll J, Daveluy S. A Review of Vitamin B12 in Dermatology. *American Journal of Clinical Dermatology*. 2015.

Buonocore D, Lazzeretti A, Tocabens P, Nobile V, Cestone E, Santin G, et al. Resveratrol-procyanidin blend: Nutraceutical and antiaging efficacy evaluated in a placebo-controlled, double-blind study. *Clin Cosmet Investig Dermatol*. 2012.

Burris J, Rietkerk W, Woolf K. Relationships of self-reported dietary factors and perceived acne severity in a cohort of New York young adults. *J Acad Nutr Diet*. 2014.

Cardona F, Andrés-Lacueva C, Tulipani S, Tinahones FJ, Queipo-Ortuño MI. Benefits of polyphenols on gut microbiota and implications in human health. *Journal of Nutritional Biochemistry*. 2013.

Chen AC, Martin AJ, Dalziell RA, Halliday GM, Damian DL. Oral nicotinamide reduces transepidermal water loss: a randomised controlled trial. *Br J Dermatol*. 2016.

Chiu A, Chon SY, Kimball AB. The response of skin disease to stress: Changes in the severity of acne vulgaris as affected by examination stress. *Arch Dermatol*. 2003.

Christensen K, Iachina M, Rexbye H, Tomassini C, Frederiksen H, McGue M, et al. Looking old for your age: Genetics and mortality [1]. *Epidemiology*. 2004.

Christensen R, Heitmann BL, Winther Andersen K, Haagen Nielsen O, Bek Sørensen S, Jawhara M, Bygum A, Hvid L, Grauslund J, Wied J, Glerup H, Fredbe U and Andersen V. Impact of red and processed meat and fibre intake on treatment outcomes among patients with chronic inflammatory diseases: protocol for a prospective cohort study of prognostic factors and personalised medicine. *BMJ Open*. 2018.

Clark AK, Haas KN, Sivamani RK. Edible plants and their influence on the gut microbiome and acne. *International Journal of Molecular Sciences*. 2017.

Collin P, Vilppula A, Luostarinen L, Holmes GKT, Kaukinen K. Review article: coeliac disease in later life must not be missed. *Alimentary Pharmacology and Therapeutics*. 2018.

Cosgrove MC, Franco OH, Granger SP, Murray PG, Mayes AE. Dietary nutrient intakes and skin-aging appearance among middle-aged American women. *Am J Clin Nutr*. 2007.

Crane JD, Macneil LG, Lally JS, Ford RJ, Bujak AL, Brar IK, et al. Exercise-stimulated interleukin-15 is controlled by AMPK and regulates skin metabolism and aging. *Aging Cell*. 2015.

Drake L. Hot sauce, wine and tomatoes cause flare-ups, survey finds. *Rosacea.org*. https://www.rosacea.org/rr/2005/fall/article_3.php. 2017.

Egawa M, Haze S, Gozu Y, Hosoi J, Onodera T, Tojo Y, et al. Evaluation of psychological stress in confined environments using salivary, skin, and facial image parameters. *Sci Rep*. 2018.

Ekiz O, Balta I, Sen BB, Dikilitas MC, Ozuguz P, Rifaioglu EN. Vitamin D status in patients with rosacea. *Cutan Ocul Toxicol*. 2014.

El-Akawi Z, Abdel-Latif N, Abdul-Razzak K. Does the plasma level of vitamins A and E affect acne condition? *Clin Exp Dermatol*. 2006.

Farrar MD, Nicolaou A, Clarke KA, Mason S, Massey KA, Dew TP, et al. A randomized controlled trial of green tea catechins in protection against ultraviolet radiation-induced cutaneous inflammation. *Am J Clin Nutr*. 2015.

Flament F, Bourokba N, Nouveau S, Li J, Charbonneau A. A severe chronic outdoor urban pollution alters some facial aging signs in Chinese women. A tale of two cities. *Int J Cosmet Sci*. 2018.

França K, Lotti T. Mindfulness within psychological interventions for the treatment of dermatologic diseases. *Dermatologic Therapy*. 2017.

Fuchs J, Kern H. Modulation of UV-light-induced skin inflammation by D-alpha-tocopherol and L-ascorbic acid: A clinical study using solar simulated radiation. *Free Radic Biol Med*. 1998.

Fujii M, Ohyanagi C, Kawaguchi N, Matsuda H, Miyamoto Y, Ohya S, et al. Eicosapentaenoic acid ethyl ester ameliorates atopic dermatitis-like symptoms in special diet-fed hairless mice, partly by restoring covalently bound ceramides in the stratum corneum. *Experimental Dermatology*. 2018.

Fukagawa S, Haramizu S, Sasaoka S, Yasuda Y, Tsujimura H, Murase T. Coffee polyphenols extracted from green coffee beans improve skin properties and microcirculatory function. *Biosci Biotechnol Biochem*. 2017.

Garg A, Chren MM, Sands LP, Matsui MS, Marenus KD, Feingold KR, et al. Psychological stress perturbs epidermal barrier homeostasis. Implications for the Pathogenesis of Stress-associated Skin Disorders. *Arch Dermatol*. 2001.

Ghodsi SZ, Orawa H, Zouboulis CC. Prevalence, severity, and severity risk factors of acne in high school pupils: A community-based study. *J Invest Dermatol*. 2009.

Grether-Beck S, Marini A, Jaenicke T, Stahl W, Krutmann J. Molecular evidence that oral supplementation with lycopene or lutein protects human skin against ultraviolet radiation: results from a double-blinded, placebo-controlled, crossover study. *Br J Dermatol*. 2017.

Haitz KA, Anandasabapathy N. Docosahexaenoic acid alleviates atopic dermatitis in mice by generating t regulatory cells and M2 macrophages. *Journal of Investigative Dermatology*. 2015.

Han XD, Oon HH, Goh CL. Epidemiology of post-adolescence acne and adolescence acne in Singapore: a 10-year retrospective and comparative study. *J Eur Acad Dermatology Venereol*. 2016.

Heinrich U, Gärtner C, Wiebusch M, Olaf Eichler, Sies H, Tronnier H, et al. Supplementation with beta-carotene or a similar amount of mixed carotenoids protects humans from UV-induced erythema. *J Nutr*. 2003.

Holowatz LA, Thompson CS, Kenney WL. L-Arginine supplementation or arginase inhibition augments reflex cutaneous vasodilatation in aged human skin. *J Physiol*. 2006.

Hon KLE, Tsang YC, Poon TCW, Pong NHH, Luk NM, Leung TNH, et al. Dairy and nondairy beverage consumption for childhood atopic eczema: What health advice to give? *Clin Exp Dermatol*. 2016.

Horrobin DF. Low prevalences of coronary heart disease (CHD), psoriasis, asthma and rheumatoid arthritis in Eskimos: Are they caused by high dietary intake of eicosapentaenoic acid (EPA), a genetic variation of essential fatty acid (EFA) metabolism or a combination of. *Med Hypotheses*. 1987.

Hwang C, Ross V, Mahadevan U. Micronutrient deficiencies in inflammatory bowel disease: From A to zinc. *Inflammatory Bowel Diseases*. 2012.

Isami F, West BJ, Nakajima S, Yamagishi SI. Association of advanced glycation end products, evaluated by skin autofluorescence, with lifestyle habits in a general Japanese population. *J Int Med Res*. 2018.

Ito N, Seki S, Ueda F. The protective role of astaxanthin for UV-induced skin deterioration in healthy people – a randomized, double-blind, placebo-controlled trial. *Nutrients*. 2018.

Jobeili L, Rousselle P, Béal D, Blouin E, Roussel AM, Damour O, et al. Selenium preserves keratinocyte stemness and delays senescence by maintaining epidermal adhesion. *Aging* (Albany NY). 2017.

Jung JY, Kwon HH, Hong JS, Yoon JY, Park MS, Jang MY, et al. Effect of dietary supplementation with omega-3 fatty acid and gamma-linolenic acid on acne vulgaris: A randomised, double-blind, controlled trial. *Acta Derm Venereol*. 2014.

Kaczmarski M, Cudowska B, Sawicka-Żukowska M, Bobrus-Chociej A. Supplementation with long chain polyunsaturated fatty acids in treatment of atopic dermatitis in children. *Postępy dermatologii i Alergol*. 2013.

Kannan R, Ng MJM. Cutaneous lesions and vitamin B12 deficiency: An often-forgotten link. *Can Fam Physician*. 2008.

Karadag AS, Tutal E, Ertugrul DT, Akin KO. Effect of isotretinoin treatment on plasma holotranscobalamin, vitamin B12, folic acid, and homocysteine levels: non-controlled study. *Int J Dermatol*. 2011.

Keen M, Hassan I. Vitamin E in dermatology. *Indian Dermatol Online J*. 2016.

Kennedy C. Mindfulness and dermatology. *International Journal of Dermatology*. 2016.

Kennedy C, Bastiaens MT, Bajdik CD, Willemze R, Westendorp RGJ, Bouwes Bavinck JN. Effect of smoking and sun on the aging skin. *J Invest Dermatol*. 2003.

Khandalavala BN, Nirmalraj MC. Rapid partial repigmentation of vitiligo in a young female adult with a gluten-free diet. *Case Rep Dermatol*. 2014.

Kim H, Moon SY, Sohn MY, Lee WJ. Insulin-like growth factor-1 increases the expression of inflammatory biomarkers and sebum production in cultured sebocytes. *Ann Dermatol*. 2017.

Kober MM, Bowe WP. The effect of probiotics on immune regulation, acne, and photoaging. *Int J Women's Dermatology*. 2015.

Krystyna RG, Elzbieta KS, Wozniak M, Zegarska B. The possible role of diet in the pathogenesis of adult female acne. *Postepy Dermatologii i Alergologii*. 2016.

Li M, Vierkötter A, Schikowski T, Hüls A, Ding A, Matsui MS, et al. Epidemiological evidence that indoor air pollution from cooking with solid fuels accelerates skin aging in Chinese women. *J Dermatol Sci*. 2014.

Lopez TE, Pham HM, Nguyen B V., Tahmasian Y, Ramsden S, Coskun V, et al. Green tea polyphenols require the mitochondrial iron transporter, mitoferrin, for lifespan extension in Drosophila melanogaster. *Arch Insect Biochem Physiol*. 2016.

Marrot L. Pollution and Sun Exposure: a Deleterious Synergy. Mechanisms and Opportunities for Skin Protection. *Curr Med Chem*. 2017.

Martin-Romero FJ, Kryukov GV, Lobanov AV, Carlson BA, Lee BJ, Gladyshev VN, et al. Selenium metabolism in Drosophila. Selenoproteins, selenoprotein mRNA expression, fertility, and mortality. *J Biol Chem*. 2001.

Meesters A, den Bosch-Meevissen YMCI, Weijzen CAH, Buurman WA, Losen M, Schepers J, et al. The effect of Mindfulness-Based Stress Reduction on wound healing: a preliminary study. *J Behav Med*. 2018.

Meinke MC, Darvin ME, Vollert H, Lademann J. Bioavailability of natural carotenoids in human skin compared to blood. *Eur J Pharm Biopharm*. 2010.

Mekic S, Jacobs LC, Hamer MA, Ikram MA, Schoufour JD, Gunn DA, et al. A healthy diet in women is associated with less facial wrinkles in a large Dutch population-based cohort. *J Am Acad Dermatol*. 2018.

Melnik BC. Linking diet to acne metabolomics, inflammation, and comedogenesis: An update. *Clinical, Cosmetic and Investigational Dermatology*. 2015.

Michaelsson G. Decreased concentration of selenium in whole blood and plasma in acne vulgaris. *Acta Dermato-Venereologica*. 1990.

Milani M, Sparavigna A. Antiaging efficacy of melatonin-based day and night creams: A randomized, split-face, assessor-blinded proof-of-concept trial. *Clin Cosmet Investig Dermatol*. 2018.

Millsop JW, Bhatia BK, Debbaneh M, Koo J, Liao W. Diet and psoriasis, part III: Role of nutritional supplements. *Journal of the American Academy of Dermatology*. 2014.

Mirnezami M, Rahimi H. Serum Zinc Level in Vitiligo: A Case-control Study. *Indian J Dermatol*. 2018.

Montgomery K, Norman P, Messenger AG, Thompson AR. The importance of mindfulness in psychosocial distress and quality of life in dermatology patients. *Br J Dermatol*. 2016.

Moseley H, Cameron H, MacLeod T, Clark C, Dawe R, Ferguson J. New sunscreens confer improved protection for photosensitive patients in the blue light region. *Br J Dermatol*. 2001.

Mostafa WZ, Hegazy RA. Vitamin D and the skin: Focus on a complex relationship: A review. *Journal of Advanced Research*. 2013.

Muizzuddin N, Matsui MS, Marenus KD, Maes DH. Impact of stress of marital dissolution on skin barrier recovery: Tape stripping and measurement of transepidermal water loss (TEWL). *Ski Res Technol*. 2003.

Mukai K, Nishimura M, Nagano A, Tanaka K, Niki E. Kinetic study of the reaction of vitamin C derivatives with tocopheroxyl (vitamin E radical) and substituted phenoxyl radicals in solution. *BBA — Gen Subj*. 1989.

NazIroğlu M, YIldIz K, Tamtürk B, Erturan I, Flors-Arce M. Selenium and psoriasis. *Biological Trace Element Research*. 2012.

Neukam K, De Spirt S, Stahl W, Bejot M, Maurette JM, Tronnier H, et al. Supplementation of flaxseed oil diminishes skin sensitivity and improves skin barrier function and condition. *Skin Pharmacol Physiol*. 2011.

Niren NM. Pharmacologic doses of nicotinamide in the treatment of inflammatory skin conditions: A review. *Cutis*. 2006.

Norström F, Sandström O, Lindholm L, Ivarsson A. A gluten-free diet effectively reduces symptoms and health care consumption in a Swedish celiac disease population. *BMC Gastroenterol*. 2012.

Nouveau-Richard S, Yang Z, Mac-Mary S, Li L, Bastien P, Tardy I, et al. Skin ageing: A comparison between Chinese and European populations: A pilot study. *J Dermatol Sci*. 2005.

Ogawa Y, Kinoshita M, Shimada S, Kawamura T. Zinc and Skin Disorders. *Nutrients*. 2018.

Öhlund I, Lind T, Hernell O, Silfverdal SA, Äkeson PK. Increased Vitamin D intake differentiated according to skin color is needed to meet requirements in young Swedish children during winter: A double-blind randomized clinical trial. *Am J Clin Nutr*. 2017.

Pappas A, Liakou A, Zouboulis CC. Nutrition and skin. *Rev Endocr Metab Disord*. 2016. Sep;17(3):443–448.

Park BW, Ha JM, Cho EB, Jin JK, Park EJ, Park HR, Kang HJ, Ko SH, Kim KH, Kim KJ. A Study on Vitamin D and Cathelicidin Status in Patients with Rosacea: Serum Level and Tissue Expression. *Ann Dermatol*. 2018.

Park SY, Byun EJ, Lee JD, Kim S, Kim HS. Air pollution, autophagy, and skin aging: Impact of particulate matter (PM10) on human dermal fibroblasts. *Int J Mol Sci*. 2018.

Parrado C, Philips N, Gilaberte Y, Juarranz A, González S. Oral Photoprotection: Effective Agents and Potential Candidates. *Front Med*. 2018.

Passarino G, De Rango F, Montesanto A. Human longevity: Genetics or Lifestyle? It takes two to tango. *Immunity and Ageing*. 2016.

Peng F, Xue C-H, Hwang SK, Li W-H, Chen Z, Zhang J-Z. Exposure to fine particulate matter associated with senile lentigo in Chinese women: a cross-sectional study. *J Eur Acad Dermatology Venereol*. 2017.

Pereira Duquia R, Da Silva Dos Santos I, De Almeida H, Martins Souza PR, De Avelar Breunig J, Zouboulis CC. Epidemiology of Acne Vulgaris in 18-Year-Old Male Army Conscripts in a South Brazilian City. *Dermatology*. 2017.

Pérez-Jiménez J, Neveu V, Vos F, Scalbert A. Identification of the 100 richest dietary sources of polyphenols: An application of the Phenol-Explorer database. *Eur J Clin Nutr*. 2010.

Picardo M, Ottaviani M. Skin microbiome and skin disease: the example of rosacea. *J Clin Gastroenterol* [Internet]. 2014;48 Suppl 1(December):S85-6. Available from: http://ovidsp.ovid.com/ovidweb.cgi?T=JS&PAGE=reference&D=mesx&NEWS=N&AN=25291137.

Puizina-Ivić N. Skin aging. *Acta dermatovenerologica Alpina, Pannonica, Adriat*. 2008.

Pullar JM, Carr AC, Vissers MCM. The roles of vitamin C in skin health. *Nutrients*. 2017.

Purba M br, Kouris-Blazos A, Wattanapenpaiboon N, Lukito W, Rothenberg EM, Steen BC, et al. Skin wrinkling: Can food make a difference? *J Am Coll Nutr*. 2001.

Ramalingum N, Mahomoodally MF. The therapeutic potential of medicinal foods. *Advances in Pharmacological Sciences*. 2014.

Roberts WE. Pollution as a risk factor for the development of melasma and other skin disorders of facial hyperpigmentation — is there a case to be made? *J Drugs Dermatol* [Internet]. 2015;14(4):337–41. Available from: http://europepmc.org/abstract/med/25844605.

Rodrigo L, Beteta-Gorriti V, Alvarez N, de Castro CG, de Dios A, Palacios L, et al. Cutaneous and mucosal manifestations associated with celiac disease. *Nutrients*. 2018.

Rodríguez-García C, González-Hernández S, Pérez-Robayna N, Guimerá F, Fagundo E, Sánchez R. Repigmentation of vitiligo lesions in a child with celiac disease after a gluten-free diet. *Pediatr Dermatol*. 2011.

Safdar A, Bourgeois JM, Ogborn DI, Little JP, Hettinga BP, Akhtar M, et al. Endurance exercise rescues progeroid aging and induces systemic mitochondrial rejuvenation in mtDNA mutator mice. *Proc Natl Acad Sci*. 2011.

Salem I, Ramser A, Isham N, Ghannoum MA. The gut microbiome as a major regulator of the gut-skin axis. *Frontiers in Microbiology*. 2018.

Sánchez-Armendáriz K, García-Gil A, Romero CA, Contreras-Ruiz J, Karam-Orante M, Balcazar-Antonio D, et al. Oral vitamin D3 5000 IU/day as an adjuvant in the treatment of atopic dermatitis: a randomized control trial. *Int J Dermatol*. 2018.

Schäfer T, Nienhaus A, Vieluf D, Berger J, Ring J. Epidemiology of acne in the general population: The risk of smoking. *Br J Dermatol*. 2001.

Schwarz A, Bruhs A, Schwarz T. The Short-Chain Fatty Acid Sodium Butyrate Functions as a Regulator of the Skin Immune System. *J Invest Dermatol*. 2017.

Schwartz RA, Janusz CA, Janniger CK. Seborrheic Dermatitis: An Overview — American Family Physician. *American Family Physician*. 2006.

Sies H, Stahl W. Carotenoids and UV protection. *Photochem Photobiol Sci*. 2004.

Silverberg JI, Silverberg AI, Malka E, Silverberg NB. A pilot study assessing the role of 25 hydroxy vitamin D levels in patients with vitiligo vulgaris. *J Am Acad Dermatol*. 2010.

Sondenheimer K, Krutmann J. Novel Means for Photoprotection. *Front Med*. 2018.

Sørensen LT, Toft BG, Rygaard J, Ladelund S, Paddon M, James T, et al. Effect of smoking, smoking cessation, and nicotine patch on wound dimension, vitamin C, and systemic markers of collagen metabolism. *Surgery*. 2010.

Stewart TJ, Bazergy C. Hormonal and dietary factors in acne vulgaris versus controls. *Dermatoendocrinol*. 2018.

Subedi L, Lee TH, Wahedi HM, Baek SH, Kim SY. Resveratrol-Enriched Rice Attenuates UVB-ROS-Induced Skin Aging via Downregulation of Inflammatory Cascades. *Oxid Med Cell Longev*. 2017.

Sun H, Chen Y, Cheng M, Zhang X, Zheng X, Zhang Z. The modulatory effect of polyphenols from green tea, oolong tea and black tea on human intestinal microbiota in vitro. *J Food Sci Technol*. 2018.

Talbott W, Duffy N. Complementary and Alternative Medicine for Psoriasis: What the Dermatologist Needs to Know. *American Journal of Clinical Dermatology*. 2015.

Tan SP, Brown SB, Griffiths CEM, Weller RB, Gibbs NK. Feeding filaggrin: Effects of l-histidine supplementation in atopic dermatitis. *Clin Cosmet Investig Dermatol*. 2017.

Tominaga K, Hongo N, Fujishita M, Takahashi Y, Adachi Y. Protective effects of astaxanthin on skin deterioration. *J Clin Biochem Nutr*. 2017.

Vandersee S, Beyer M, Lademann J, Darvin ME. Blue-violet light irradiation dose dependently decreases carotenoids in human skin, which indicates the generation of free radicals. *Oxid Med Cell Longev*. 2015.

Vierkötter A, Hüls A, Yamamoto A, Stolz S, Krämer U, Matsui MS, et al. Extrinsic skin ageing in German, Chinese and Japanese women manifests differently in all three groups depending on ethnic background, age and anatomical site. *J Dermatol Sci*. 2016.

Vollmer DL, West VA, Lephart ED. Enhancing skin health: By oral administration of natural compounds and minerals with implications to the dermal microbiome. *International Journal of Molecular Sciences*. 2018.

Wacewicz M, Socha K, Soroczyńska J, Niczyporuk M, Aleksiejczuk P, Ostrowska J, et al. Selenium, zinc, copper, Cu/Zn ratio and total antioxidant status in the serum of vitiligo patients treated by narrow-band ultraviolet-B phototherapy. *J Dermatolog Treat*. 2018.

Weisshof R, Chermesh I. Micronutrient deficiencies in inflammatory bowel disease. *Current Opinion in Clinical Nutrition and Metabolic Care*. 2015.

Willcox DC, Scapagnini G, Willcox BJ. Healthy aging diets other than the Mediterranean: A focus on the Okinawan diet. *Mech Ageing Dev*. 2014.

Williams S, Krueger N, Davids M, Kraus D, Kerscher M. Effect of fluid intake on skin physiology: Distinct differences between drinking mineral water and tap water. *Int J Cosmet Sci*. 2007.

Xu X, Fu W-W, Wu W-Y. Serum 25-Hydroxyvitamin D Deficiency in Chinese Patients with Vitiligo: A Case-Control Study. *PLoS One*. 2012.

Yamada T, Alpers DH et al. *Textbook of gastroenterology*. Chichester, West Sussex: Blackwell Pub; 2009.

Yildizgören MT, Tofral AK. Preliminary evidence for vitamin D deficiency in nodulocystic acne. *Dermatoendocrinol*. 2014.

Yin L, Morita A, Tsuji T. Skin premature aging induced by tobacco smoking: The objective evidence of skin replica analysis. *J Dermatol Sci*. 2001.

Yoshida S, Yasutomo K, Watanabe T. Treatment with DHA/EPA ameliorates atopic dermatitis-like skin disease by blocking LTB4 production. *J Med Invest*. 2016.

Yousefzadeh MJ, Zhu Y, McGowan SJ, Angelini L, Fuhrmann-Stroissnigg H, Xu M, et al. Fisetin is a senotherapeutic that extends health and lifespan. *EBioMedicine*. 2018.

Zhou F, Huang X, Pan Y, Cao D, Liu C, Liu Y, et al. Resveratrol protects HaCaT cells from ultraviolet B-induced photoaging via upregulation of HSP27 and modulation of mitochondrial caspase-dependent apoptotic pathway. *Biochem Biophys Res Commun*. 2018.

Ziboh VA, Chapkin RS. Biologic significance of polyunsaturated fatty acids in the skin. *Arch Dermatol*. 1987.

Index

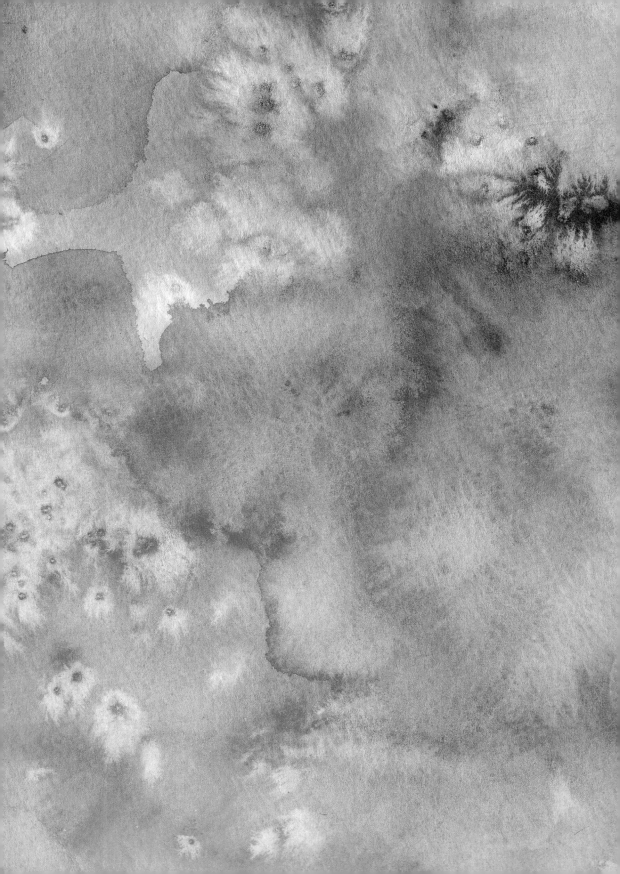